Wisdom in the Age of Intelligent Machines

Edward H. Spence

Wisdom in the Age of Intelligent Machines

palgrave
macmillan

Edward H. Spence
School of Humanities
University of Sydney
NSW, Australia

ISBN 978-3-031-93882-5 ISBN 978-3-031-93883-2 (eBook)
https://doi.org/10.1007/978-3-031-93883-2

This Palgrave Macmillan imprint is published by the registered company Springer Nature Switzerland AG.
The registered company address is: Gewerbestrasse 11, 6330 Cham, Switzerland

We shall not cease from exploration, and the end of all our exploring
will be to arrive where we started and know the place
for the first time.
—T.S. Eliot, Four Quartets, Little Gidding, V.

To Sophia, Kaye and Diotima

Acknowledgements

I would like to express my heartfelt thanks and gratitude to the following people who in various ways have helped me bring this book to fruition. The acquisition editor Amy Invernizzi, Senior Editor, Philosophy, Palgrave Macmillan, Springer Nature, for her constant support throughout the publication process of this book. My mentor, and ever-present Alan Gewirth, Kaye Spence for her dedicated work in supporting me to finish this book, my friends and colleagues John Weckert, Seumas Miller, Rick Benitez, Michael Boylan, David Field, Richard Spence and Clare Donnely, for their useful comments and discussions on aspects of this book and all my colleagues from the 4TU Centre for Ethics and Technology, as well as my long-standing colleagues, Luciano Floridi, Charles Ess, Johnny Soraker, Jeroen van den Hoven, Judith Simmons, Peter Paul Verbeek. Ibo van de Poel and Philip Brey, and not least Plato, Socrates, Zeno of Citium, Marcus Aurelius and especially Diotima.

Contents

About the Author

Edward H. Spence is a philosopher and author specialising in ethics, technology and wisdom. He is an honorary affiliate of the School of Humanities, Department of Philosophy, University of Sydney and Charles Sturt University, and a senior research fellow at the 4TU Centre for Ethics and Technology in the Netherlands.

He has authored over 100 peer-reviewed publications and books, including *Wisdom in the Age of Intelligent Machines* (forthcoming), *Stoic Philosophy and the Control Problem of AI Technology* (2021), and *Media Corruption in the Age of Information* (2021). His research explores AI ethics, media integrity and the digital age's moral challenges.

Dr. Spence is the founder of the **Philosophy Plays Project**, using theatre to engage audiences in philosophical discussions. His plays—performed at the Greek Festival of Sydney and international arts festivals—examine wisdom, AI and Stoic thought.

As President of the **Zeno Stoic Centre**, he promotes Stoic philosophy worldwide. Through his books, research and performances, Dr. Spence encourages ethical reflection on the digital age, bridging ancient wisdom with contemporary challenges.

CHAPTER 1

Introduction

Wisdom, the Peculiar Knowledge of Humans, Gods and Machines

That which we are, we are. That is the wisdom of great age, the acceptance of our impermanence while drawing consolation from the memories of what we did and what, with luck, we might yet do. Also, perhaps, we can hope that our contribution will not be entirely forgotten as wisdom and understanding spread outwards from the Earth to embrace the cosmos.
—James Lovelock (2020, 130)

Starting with a brief overall global view of wisdom, the peculiar knowledge, shared by gods and humans stretching through ancient times to the present from East to West incorporating sacred texts from Judaism, Hinduism, Buddhism, Taoism, Islam and other Indigenous religions and cultures including of course the ancient religion and myths of Rome and Greece, the book's main focus is the examination and normative evaluation of the impact AI technology has on our well-being now and potentially in the future. The key question is what is technology good for and specifically artificial intelligence (AI) technology? This book is motivated by the need to answer that question by examining the impact AI technology (henceforth technology) has on human well-being, through the notion of Wisdom. For in the absence of any tangible or potential *eudaimonic* benefit for society on our collective well-being, what is technology

E. H. Spence, *Wisdom in the Age of Intelligent Machines*,
https://doi.org/10.1007/978-3-031-93883-2_1

good for, especially in cases where the eudaimonic impact of technology on society is not only very negligible but harmful?

To answer that question, the book develops and applies a meta-normative model of wisdom whose primary purpose is to determine the essential conditions that any normative theory that seeks to assess the impact of technology on well-being must adequately address in order to be able to account for, explain and evaluate the *capability* of any technological product or process in its design and/or its use to contribute in some way, if any, to the good life for the *well-being* of individuals and society generally. Using a theoretical normative tripartite model of Wisdom, the *Dual Obligation Information Theory-Wisdom (DOIT-Wisdom)* (Spence 2011) that comprises three inter-related normative components, the *epistemic*, *ethical* and the *eudaimonic*, within an overarching conceptual framework, this book examines the actual and potential impact of AI technology on well-being. For example, information and communication technology (ICT) and AI technologies that do not protect the justified autonomy and privacy rights of its users or allow and do not restrict the indiscriminate dissemination of misinformation and disinformation, as well as the indiscriminate harvesting of user's information for the primary gain of Big Tech companies such as Facebook and Google, would have a low and negative capability for well-being. As such, it would potentially generate a negative epistemic, ethical and eudaimonic impact for its users and generally, a negative impact on society and the global community overall and therefore would not be wise to use such technologies.

This book offers an innovative approach to evaluating information and knowledge and its relation to the good life, in the age of Autonomous intelligent machines, through the concept of Wisdom. *Wisdom* understood as a type of *meta-information* and *meta-knowledge*, which comprises epistemic, ethical and eudaimonic features, providing a direct conceptual and practical link between the concepts of information, knowledge, the good life and well-being. More generally, it provides a direct link between technology and in particular information and communication technology (ICT) and artificial intelligence (AI) technologies, and the good life and well-being. As such, the concept of wisdom allows for a direct *normative evaluation* of the impact of the dissemination of information through ICTs and AI technologies on the well-being of individuals and society at large. It provides a theoretical rationale to demonstrate the important and relevant role that wisdom plays in the specific evaluation of *information* in the digital and technological age in which we now live. To that end, a methodological approach is used to show how some different general *types* of practical manifestations of digital information can be normatively evaluated (if they are good or bad for us) through the application of the concept of wisdom. In sum, this book seeks to show that although related, wisdom is conceptually distinct and conceptually different from information, and knowledge and by implication intelligence in a crucial way. For wisdom unlike information and knowledge provides a person with *understanding* concerning the *techne viou* or *craftsmanship of living* in the sense of *knowing how* to *evaluate and apply* relevant information or knowledge for living a good life for the attainment of eudaimonia or happiness, and in addition, an appreciation in knowing *why* such a life constitutes a good life. For the Stoics, and generally for the Hellenistic philosophers, including the Epicureans, the Cynics and the Sceptics, philosophy was conceived as a way of life or *techne viou*, the art of living a good life through virtue, as the Stoics, for example, maintained, for the attainment of happiness. For one could be intelligent, possess information and knowledge, be wealthy and powerful, yet not know how to wisely apply that knowledge in living a good life for the attainment of well-being or happiness for themselves and others.

Chapter Synopsys

Chapter 2. Wisdom and Well-being: A Wisdom Theory Model

This chapter provides a general philosophical groundwork for the theoretical and applied normative evaluation of information generally and digital information specifically in relation to the good life for the attainment of well-being, through the overarching notion of wisdom on the basis of a theoretical normative tripartite model of Wisdom, the *Dual Obligation Information Theory-Wisdom (DOIT-Wisdom)* (Spence 2011) that comprises three inter-related normative components, the *epistemic*, *ethical* and the *eudaimonic*, within the overarching conceptual framework. Following from the question posed in in the Introduction (Chap. 1) what is technology good for and specifically AI technology, the overall aim of this chapter is to address more specifically, the question of how the epistemology and ethics of information and communication technologies (ICTs), as well as the epistemology and ethics of AI technologies (abbreviated to t*echnology*, henceforth) can be expanded to include more centrally the issue of its *eudaimonic* impact of technology on well-being for individuals, and society. To answer that question, the chapter provides by way of a theoretical groundwork, the concept of *wisdom* understood as a type of *meta-knowledge* as well as a type of *meta-virtue*, which can enable us as individuals and collectively as a society, and humanity generally, to both know in principle what constitutes a good life, and how to successfully apply that knowledge as *know-how* in living such a life in practice, for the ultimate purpose of attaining well-being or eudaimonia. This answer will be based on the main argument presented in this chapter that the notion of *wisdom* understood as a triadic concept, at once a *meta-epistemological*, *meta-ethical* and *meta-eudemonic*, provides the essential conceptual link between information on the one hand and well-being on the other. In the digital age of information, both the theoretical examination and analysis of the question of how information, in its technological construction and dissemination, relates to well-being, and the provision of an adequate answer to that question are essential for developing a deeper understanding of how to evaluate the theoretical and practical implications and impact of digital information on well-being for individuals and societies.

Chapter 3. Wisdom and Control: Applying Stoic Philosophy to Technology

> O my friend, why do you who are a citizen of the great and mighty and wise city of Athens, care so much about laying up the greatest amount of money and honour and reputation, and so little about wisdom and truth ... Are you not ashamed of this? (Plato, *Apology*, 1952)

This chapter examines how Stoic and neo-Stoic philosophy as a philosophy of practical and applied wisdom applies to technology, and specifically to Information Communication Technologies (ICTs) and Artificial Intelligence (AI) technologies. Unless otherwise stated for convenience I will refer to these technologies collectively, as "technology." The chapter comprises three sections: (1) Comparison of the control problem of AI technology and how it aligns to the control problem of Stoic philosophy; (2) the core principles of Stoic philosophy and how they apply to technology; and (3) why Stoic philosophy as practical wisdom is of relevance and importance to technology.

Let me begin by clarifying, that the core problem of technology is not technology as such, but the use of technology by the Big Tech companies, such as Facebook and Google, for their primary financial gain to the detriment of the users' well-being and society generally. The recent scandal of Cambridge Analytica and Facebook involving breach of privacy and lack of consent of users' data is a case in point. That case and its normative implications and impact for society will be examined in Chap. 5.

Given that any technology cannot be good for itself or the good of the Big Tech companies that manage them, but good for humanity overall, any instrumental positive value of such technologies must not be at the cost of their negative eudaimonic impact on society. Ultimately, the core problem of technology is not technical to be solved by engineers but a moral problem to be solved collectively by society and the global community.

The overarching rationale for the application of Stoic and neo-Stoic philosophy to evaluate the impact of technology on society is that Stoic philosophy as a way of life with its focus on the attainment of eudaimonia or well-being is conceptually and practically relevant and important to the eudaimonic assessment of technology. The Dual Obligation Information-Wisdom Theory (DOIT-Wisdom) designed to enable the operationalisation of the application of Stoic ethics to technology forms a central feature of the application of Stoic philosophy to technology.

Chapter 4. Artificial Agency and Moral Autonomy: The Quest for Artificial Wisdom

This chapter will address a very important topic in the Philosophy and Ethics of Technology, which is that of *artificial mediated agency and autonomy* and its moral significance and impact on human well-being. Furthermore, it will address the key question of this chapter whether artificial agency and autonomy can be *extended to the notion of wisdom*: if agency is mediated and distributed between humans and intelligent machines, can there also be a mediated wisdom? Although several well-known philosophers of technology, such as Luciano Floridi (2019, 2014), Peter-Paul Verbeek (2011, 2014), Richard Heersmink (2017), to mention but a few, have researched and written on the issue of *artificial moral mediation*, very little research has been published on the topic of wisdom with regard to AI technology and its impact on well-being. This chapter and the book in general, sets out to address and fill that gap. According to Verbeek, "only by recognizing this interweaving of human and technologies can we take responsibility for the ways in which *technologies have an impact on society and on human existence* [my emphasis] – in practices of technology design, implementation, and use" (2014). In a clear summary Heersmink summarises Verbeek's distributed moral agency (2017) thus:

> By drawing on actor-network theory, Verbeek (2011) develops the notion of "distributed moral agency." He specifically focuses on the moral aspects of obstetric ultrasound imaging, an imaging technology used to visualise a foetus in the mother's womb, which is usually done a week twelve and twenty of pregnancy. This technology does not just provide a neutral and transparent window to the womb but has several morally important aspects…so obstetric ultrasound imaging mediates and transforms the relationship parents have with their foetus, providing them with information on the health of their foetus that is important for moral decision-making regarding pregnancy and abortion. For these reasons, moral agency should be seen as distributed across humans and technology. In Verbeek's words: Ultrasound imaging actively contributes to the coming about of moral actions and the considerations behind moral actions. The example therefore shows that moral agency should not be seen as an exclusive human property; it is distributed among human beings and nonhuman entities. (2011)

However, Verbeek, Floridi and Heersmink and the other aforementioned philosophers, do not refer to the important role that wisdom should play in recognising and assessing the mediated impact of technology on well-being, and it is precisely that missing inclusion of wisdom and its gap that this chapter and this book overall, intends to fill. Namely, to examine and evaluate the impact of mediated AI technologies on our well-being, in terms of our agency, autonomy, morality, and most importantly as we shall examine in Chap. 8 of this book, our dignity, and do so through the concept of wisdom. One key question this chapter will examine is whether there can be mediated wisdom along the lines that Floridi, Verbeek and Heersmink apply to mediated moral agency or is some other way that retains the important role of human autonomy and dignity, which is of central importance to human agency, a topic that is examined in Chap. 8 of this book.

In an important innovative paper "Artificial wisdom: a philosophical framework, Cheng-hung Tsai examines the notion of artificial wisdom (2020). In his paper he explains

> why artificial wisdom matters and how artificial wisdom is possible (in principle and in practice) .... the result is a conceptual framework that guides future research on creating artificial wisdom. (2020)

Contrary to that view, Jason Swartwood and Valerie Tiberius (2019) argue:

A wise person's understanding is *uncodifiable:* it is possible to derive good guidance in any particular situation. Morality is too complex, virtue ethicists argue, to capture with such a set of principles. For instance, someone might try to defend the following moral principle: An action is right when, of all the actions possible in a situation, it could be expected to produce the most total long-term happiness when the effects on everyone are added up. If this principle accurately described what matters in all situations, then a non-virtuous person could use it to derive (after suitable inquiry into the likely consequences of different actions) reliable guidance about what to do. Virtue Ethicists argue that principles like this do not succeed in codifying wise understanding.

Julia Annas provides additional arguments against codifiability (Annas 2004). If Virtue Ethicists are correct that wise understanding is uncodifiable, then that would imply that a reliable understanding of how to conduct oneself must be acquired through some sort of experience and reflection .... Aristotle thought we need reflection that focuses on both the universal and the particulars of situations (NE 1141b10-15, 1142a14).

Swartwood and Tiberius raise an important characteristic of wisdom that resists codifiability, that is, wisdom is a key characteristic of Virtue Ethics and by extension Stoic philosophy and Hellenistic philosophy generally, which is that wisdom is a way of life for the attainment of eudaimonia as argued in Chap. 3 of this book, and thus specific to individual lives. This chapter will examine in further detail whether even if uncodifiable, wisdom, however, could be mediated in some way to be defined, through AI technology as suggested with regard to morality by Verbeek and Floridi, whilst at the same time, preserving the moral agency of wise human beings that remains under their full control.

Chapter 5. The Corruption of Information: Tech, Media and Truth in the Age of AI

> The fault dear Brutus is not in our stars but in ourselves.
> (Shakespeare, Julius Caesar, Act I, Scene III, L. 140–141)

Choosing *Facebook* as an illustrative example of what is referred to in this chapter as the *6th Estate* in order to distinguish it conceptually from the 4th and 5th Estates, the primary objective of the first section of the chapter is to examine if digital information created, disseminated and mediated increasingly via the Big Tech companies such as Facebook and Google, are also subject to those same normative principles as the 4th and 5th Estates. Having established that Facebook, as an illustrative example, is in essence a *media company*, the second section of the chapter will demonstrate, on the basis of the *Dual Obligation Information Theory (DOIT)* and the DOIT-Wisdom model examined in Chap. 5, that Facebook is also subject to the same normative principles and requirements that other media companies of the 4th and 5th Estates, are. The third section of the chapter will examine if Facebook's role in the Cambridge Analytica case not only violated the fundamental normative principles and requirements to which all media companies are subject, but moreover, its actions constituted media corruption. More generally, Facebook's media corruption as illustrated in the Cambridge Analytica case, relates to a conflict of interest emanating from its Business Model that in its design and practice is conducive to systemic media corruption.

Chapter 6. Wisdom and Intelligent Machines: Who Is in Control?

> *The ultimate form of control is individual sovereignty, understood as self-ownership, especially one's own body, choices, and data. The fight for digital sovereignty is an epochal struggle.* (Luciano Floridi 2019)

As in Chap. 3 and emphasised in Chap. 4 with regard to mediated agency, the concept of dignity is central to both Stoic and neo-Stoic philosophy and specifically with regard to the problem of control, as expressed in Stoic philosophy and analogously as it applies to technology, and in particular, ICTs and AI technologies. As most of the key problems in the use of those technologies are generated through the use of AI algorithms, those problems will be analysed in conjunction with the normative impact those technologies have on well-being, on individuals and society generally. This chapter will examine and evaluate in more detail the impact of ICTs and AI technologies on well-being through a close analysis of two major problems in AI machine learning and neural network algorithms: that of the inscrutability or the black box problem and the control problem. This chapter will also suggest possible solutions to both of those problems, some of them proposed recently by AI researchers. A key concern that both those problems raise is that without transparency there is no accountability and without accountability there is no normative way to monitor AI systems for potential adverse and harmful impacts on society, let alone control them. As such, the common collective good of society and humanity generally, is at stake. This chapter will also introduce the notion of anticipatory normative controls through design to prevent or at least minimise the risk of the adverse impact of AI technologies on well-being through opacity or lack of control.

Can we now like the Trojans standing on the broken walls of Troy looking down at the ruins of their fallen city hear the warnings of Cassandra "beware of Big Tech bearing gifts" before we too find the last citadel of our inner freedom and autonomy are under the control of the Big Tech companies.

Chapter 7. Wise Guys and Smart Machines

Beginning with Chap. 3, this book sets out to apply the theoretical Stoic and neo-Stoic framework and its interrelated *Dual Obligation Information Theory(DOIT)-Wisdom* model in Chap. 2 to examine and normatively evaluate the normative impact of some of the practices of Big Tech companies (Apple, Amazon, Facebook, Google and Microsoft) through the operations of their ICTs and AI technologies, and specifically, in terms of their epistemic, ethical and eudaimonic impact they have on individual citizens as users and data subjects of those technologies, and collectively on society and generally the global community, through the overarching unified triadic concept of Wisdom.

Applying the analysis and argument for dignity and its conferring universal rights of freedom and well-being to users and data subjects of the ICT and AI technologies of Big Tech companies, a major issue identified under that topic, is that personalisation of subjects' data by algorithms poses a threat to their autonomy and moreover to their digital identity. This is a crucial issue, one that will be examined in terms of Stoic and neo-Stoic philosophy with regard to technology's control over our judgements and decision-making, exercised by Big Tech companies, through the use of their persuasive technologies designed to gain control over our autonomy, at least our decision-making capacity, that is a key principle for the Stoics and New-Stoics and one that ought to be under our personal control, as our individual dignity as well as our collective eudaimonia or well-being as a society, is dependent upon it.

Yet another insidious and potentially crucial existential problem not just for data subjects but humanity generally, is the split of our personal *identity* into two separate identities: a *group digital-identity* controlled externally by AI algorithms, and our *natural individual identity* controlled internally by us as individuals, but whose autonomy, however, is also precariously under the systemic control of the Big Tech companies through their persuasive technologies enabled by their invasive algorithms.

In this chapter this identity problem is referred to as the *identity* paradox. Namely, that our individual identity is under threat by the digitalisation and datafication of our identity through opaque, incomprehensible and unaccountable algorithms that construct digital identities for us without our knowledge or consent, based on statistical data gathered at group level that has very little, if anything, in common with our own natural individual identities. Crucially, this poses a significant existential problem, not just for data subjects but humanity generally. For the split of our identity into two separate identities: a *group digital identity* controlled externally by algorithms, and our *natural individual identity* controlled internally by us as individuals, but one whose autonomy, however, is also precariously under the systemic manipulative control of the persuasive technologies of the Big Tech companies. The effects linked to automated decision-making and profiling that undermine self-determination lead systemically to *data determinism* and consequently to *statistical dehumanisation* and, moreover, to *de-individualisation* (Aizenberg and Van Den Hoven 2020) and the *collectivisation of identity* of persons, which we identified metaphorically as the *Borg problem.* As data subjects, we are normatively entitled to digital rights to our autonomy, identity and dignity, and those rights ought to be safeguarded and promoted by legislation through a shared social collective responsibility in order to ensure normative compliance. That is what wisdom requires both *protreptically* through the Virtues as in Stoic philosophy and *prescriptively* through Universal Rights in accordance with neo-Stoic philosophy as examined in Chap. 3, which combines virtue ethics and universal human rights.

The chapter will examine and argue for the adoption of a *eudaimonic Stoic and humanistic approach* to AI technologies on the basis of Wisdom towards the design and use of ICTs and AI technologies, and other emerging technologies in general, with the aim of: (a) ameliorating or at least minimising the negative and harmful impacts of technology on human well-being; (b) enhancing the design and uses of technology for the overall eudaimonic benefit of humanity presently and in the future; (c) to ensure that those technologies are not under the exclusive control of the Big Tech companies that own and operate them for their primary gain and benefit, but under the shared societal control of citizens and their authorised democratic representatives and institutions, as individuals and collectively as societies, for the common good of citizens, as well as the global community.

Chapter 8. What's Love Got to Do with It?

What's Love Got to Do with It? So goes the popular song by Tina Turner. A good song, a simple song, a song to sing along, to dance to, but what does that have to do with the present objective of this chapter concerning the control problem of AI technology? A lot as it happens. Or at least I will speculatively suggest. For *philo-sophia* was conceived by ancient Greek and Hellenistic philosophers including the Stoics, to be the *love of wisdom*. In Plato's *Symposium*, love plays a central role at once metaphysical, ethical and eudaimonic, and not least aesthetic, as the highest immutable good, a transcendent love of that which is always at once beautiful and good. Might such a notion of this higher form of Platonic love, also present in Stoic and neo-Stoic philosophy, especially through the notions of *cosmopolitanism* and *oikeiosis*, contribute if not to a solution, at least to a better theoretical normative understanding of where a solution to the control problem of AI technology might lie with regard to the relationship of humans to AI machines? Can we motivate and AI superintelligent machines to *love* us and we in turn learn to reciprocate that *love* for them?

Surprisingly given the central place that the concept of love plays in human civilisation throughout the ages, expressed through culture, music, poetry, literature, art and philosophy, as well as in most world religions, such as Christianity, Islam, Buddhism, Hinduism, Judaism among others, the notion of love though indirectly implied in concepts such as justice, solidarity, empathy, equality, compassion, has received scant direct attention in AI research. Yet it would seem at first blush a fertile concept to explore as at once a theoretical and practical concept of providing, at least conceptually, a solution to the control problem of AI technology. Especially, with regard to the goal-alignment control problem we examined earlier in this chapter, namely, how to align our goals with those of a superintelligent AI. As Max Tegmark tells us: "Figuring out how to align the goals of a superintelligent AI with our goals isn't just important, but also hard, In fact, it's currently an unsolved problem" (Tegmark 2018, 260). Stuart Russell's *beneficial machines,* that is, "machines whose actions can be expected to achieve our objectives. Because these objectives are in *us*, and *not in them*" and therefore will "defer to humans," offers, as we saw earlier a promising approach to solving the goal-alignment problem of AI and the control problem AI technology we have been examining in this chapter and generally throughout this book. However, as we also examined earlier with regard to Russell's beneficial machines solution, the problem with regard to superintelligent AI seems to still remain. For what would *motivate* a superintelligent AI to comply and defer to our human objectives and not to its own objectives? Given its great motivating power, might not love, motivate a superintelligent AI to be positively predisposed towards human beings and not harm them but always seek to do them good, and do so because it learns to loves us, in an analogous way that an omnipotent, omnipresent and benevolent God, at least according to most of the world religions "loves us." So much so, that according to Christianity, God so much-loved us, that he sacrificed his own son to save us from our sins for our ultimate redemption and salvation. The comparison here is of course only meant allegorically and metaphorically to emphasise the central role that the concept of love plays in human civilisation. Nevertheless, the comparison by allegory just as Plato's notion of love in his *Symposium,* suffice to provide, just as in ancient myths, as for example, Ovid's *Metamorphoses*, a strong *conceptual link* to a possible theoretical addition to a solution to the control problem of AI technology through the concept of love.

As Philip Larkin puts it in the last line of his poem *An Arundel Tomb*, "*What will survive of us is love.*" If we know that love exists and is experienced between humans and animals such as dogs, cats, parrots, rabbits, to mention but a few, notwithstanding the vast intellectual disparity and intelligence between us and animals, then analogously it is conceptually feasible that love as an important aspect of human lives, could be developed and be experienced, between us and Artificial General Intelligence (AGI) machines in the form of say, embodied robots, such as Data, in *Star Trek*, and the Hologram Doctor in *Voyager*, for example. Conceptually at least there doesn't appear to be any rational reason why that is not feasible, though there might be technical complications and as yet unforeseen social reasons. As Tina Turner tells us in her song, "its only logical." Time will tell. What we do know, however, is that love and affection is species-neutral and transcends species as our own relationship with animals demonstrates.

If AGI machines can be designed to develop rational minds that though far more intelligent than us, are not as wise as we are, is there any rational reason, barring any unforeseen technological difficulties, why rational *true minds* cannot love and care for each other's eudaimonic good through the overarching concept of a universal love as conceived by Plato and the Stoics? I conclude, with *Sonnet 116* by William Shakespeare who also conceived of such a universal and love that permeates, as the Stoics believed, the whole of Nature. A *Love* that is at one with *Wisdom*, for as Diotima tells us, "Eros who is in love with what is beautiful, is a lover of wisdom" (Plato 1997).

References

Aizenberg, Evgeni and Van Den Hoven, Jeroen. (2020). "Designing for Human Rights in AI." *Big Data and Society.* 1(1):1-12.

Annas, Julia. (2004). "Happiness as Achievement." *Daedalus*, 133(2):44–51.

Floridi, Luciano. (2014). *The Fourth Revolution: How the Infosphere is Reshaping Human Reality.* Oxford England: Oxford University Press.

Floridi, Luciano. (2019). "The Fight for Digital Sovereignty: What It Is and Why it Matters, Especially for the EU." *Philosophy and Technology.* 33:369-78

Heersmink, Richard. (2017). "Distributed Cognition and Distributed Morality: Morality, Agency and Systems." *Science Engineering Ethics*, 23(2):431–448.

Lovelock, James. (2020) *Novacene - The Coming Age of Hyperintelligence.* Penguin Press.

Plato. (1952) The Apology of Socrates. 2nd Edition. Penguin Classics.

Plato. (1997). *Symposium and Death of Socrates.* Translated by Tom Griffith. Hertfordshire: Wordsworth Classics.

Spence, E.H. (2011) "Information, Knowledge and Wisdom: Groundwork for the Normative Evaluation of Digital Information and its Relation to the Good Life." *Ethics Inf Technol* 13, 261–275. https://doi.org/10.1007/s10676-011-9265-7

Swartwood, Jason and Tiberius, Valerie. (2019). "Philosophical Foundations of Wisdom." In T*he Cambridge Handbook of Wisdom* (Eds Robert Sternberg and Judith Glueck) Cambridge University Press.

Tegmark, Max. (2018). *Life 3.0: Being Human in the Age of Artificial Intelligence.* UK: Penguin Books.

Tsai, Cheng-hung. (2020). "Artificial Wisdom: A Philosophical Framework." *AI & Society*. DOI:10.1007/s00146-020-00949-5

Verbeek, Peter-Paul. (2011). *Moralizing Technology: Understanding and Designing the Morality of Things*. Chicago University Press.

Verbeek, Peter-Paul. (2014). *The Moral Status of Technical Artefacts*. Dordrecht: Springer.

CHAPTER 2

Wisdom and Well-Being: A Wisdom Theory Model

Introduction

Where is the wisdom we have lost in knowledge?[1]

This chapter[2] provides a general philosophical groundwork for the theoretical and applied normative evaluation of information generally and digital information specifically in relation to the good life for the attainment of well-being. As we saw in Chap. 3, a good life conceived by the Stoics as a life of virtue in accordance with nature was both a necessary and sufficient condition for the attainment of eudaimonia or well-being. Also following on from the question "What Is Technology Good For?" the overall aim of this chapter is to address more specifically the question of how the ethics of information and communication technologies (ICTs), as well as the ethics of AI technologies (abbreviated to technology, henceforth), can be expanded to include more centrally the issue of its eudaimonic impact of technology on well-being for individuals, and

[1] This line is from a poem from T. S. Eliot, Choruses from The Rock, 1934.

[2] This chapter has its conceptual origins in a three-year research project I undertook in the Department of Philosophy at the University of Twente, the Netherlands, 2006–2009, as a research fellow of a VICI international research project for the Evaluation of New Media in Relation to the Good Life. The project was funded by the National Dutch Research Council (NWO) under the supervision of Professor Philip Brey.

E. H. Spence, *Wisdom in the Age of Intelligent Machines*,
https://doi.org/10.1007/978-3-031-93883-2_2

society. To answer that question, the chapter provides by way of a theoretical groundwork, the concept of wisdom under stood as a type of meta-knowledge as well as a type of meta-virtue, which can enable us as individuals and collectively as a society, and humanity generally, to both know in principle what constitutes a good life, and how to success fully apply that knowledge as know-how in living such a life in practice, for the ultimate purpose of attaining well-being or eudaimonia. This answer will be based on the main argument presented in this chapter that the notion of wisdom understood as a triadic concept, at once a meta-epistemological, meta ethical and meta-eudemonic, provides the essential conceptual link between information on the one hand and well-being on the other. In the digital age of information, both the theoretical examination and analysis of the question of how information in its technological construction and dissemination relates to well-being, and the provision of an adequate answer to that question are essential for developing a deeper understanding of how to evaluate the theoretical and practical implications and impact of digital information on well-being for individuals and societies. In offering this theoretical groundwork the chapter will not provide a detailed examination and evaluation of specific normative issues, which arise in the production, dissemination and use of digital information in particular contexts and instances, as that will be the topic of examination in Chaps. 6, 7, and 8 of this book. This chapter will, however, provide in Section "Wisdom and the Evaluation of Digital Information in Relation to a Good Life", a methodological approach of how different types of some major practical manifestations of digital information and its related AI issues can be evaluated using the meta-theoretical framework proposed in this chapter. The chapter comprises three main inter-related sections. Section "A Universal Model for Evaluating the Normative Quality of Digital Information", provides a summary of an argument whose primary aim is to demonstrate a meta philosophical model, the *Dual Obligation Information Theory (DOIT)* (Spence 2009) to be used in the analysis and evaluation of digital information in terms of its inherent normative categories. Those categories are the epistemological and the ethical. That argument will, in turn, motivate and advance the main argument of this chapter, that the triadic notion of wisdom, at once meta-epistemic, meta-ethical and meta-eudemonic, provides the essential theoretical term that conceptually links digital information and its related AI algorithmic manifestations, to the good life and well-being.[3] This forms the overall objective of Section "Information, Knowledge and Wisdom", which is the examination and evaluation of the theoretical and practical relationship between information, knowledge and wisdom, based on the Dual Obligation Information-Wisdom Theory (DOIT-Wisdom (Spence 2011).[4] The main argument presented in this chapter is that the notion of wisdom understood as being at once a meta-epistemic, meta-ethical and meta-eudemonic concept, provides the essential conceptual link between information on the one hand and the good life in the attainment of eudaimonia or well-being on the other. In the digital age of information, both the theoretical examination and analysis of the question of how information in its technological construction and dissemination relates to the good life and the attainment of well-being, and the provision of an adequate answer to that question are essential for developing a deeper understanding of how to evaluate the theoretical and practical implications and impact of information on well-being, for individuals and societies generally. Of central importance, *Information, Knowledge and Wisdom*, which provides the focal direction of the chapter, offers an innovative approach in evaluating information and its relation to the good life, through the concept of wisdom. Wisdom understood as a type of meta-information or meta-knowledge, which comprises also essentially, epistemic, ethical and eudaimonic features, provides a direct conceptual and practical link between the concepts of information, knowledge, the good life and well-being. More generally, it provides a direct link between ICT and AI technologies and the good life and well-being. As such, the concept of wisdom allows for a direct normative evaluation of the impact of the dissemination of information through ICTs and AI technologies on the good life and well-being. Following on from Sections "A Universal Model for Evaluating the Normative Quality of Digital Information" and "Information, Knowledge and Wisdom", Section "Wisdom and the Evaluation of Digital Information in Relation to a Good Life", provides a theoretical rationale to demonstrate the important and relevant role that wisdom plays in the specific evaluation of digital information. To that

[3] Henceforth, understand the use of the concept of information, used in this chapter, to relate to digital information generally, including its related manifestations of production and dissemination via ICTs and AI technologies, including big data and algorithms.

[4] This chapter is based on an earlier version of the paper published by Spence E. 2011: Information, Knowledge and Wisdom: Groundwork for the Normative Evaluation of Digital Information and Its Relation to the Good Life, Ethics and Information Technology, (2011).

end, a methodological approach is used to show how some different general types of practical manifestations of digital information can be normatively evaluated through the application of the concept of wisdom as developed in Section "Information, Knowledge and Wisdom". That approach also helps establish the close conceptual connection between the DOIT-Wisdom theory to Stoic and neo-Stoic philosophy, of which wisdom is a core essential feature. However, before proceeding, we should first address the general question of why wisdom is at all relevant to the relationship between information and a good life? Why not proceed directly to examining the role that information itself plays in the good life and well-being? Why do we need the intermediary concept of wisdom? The answer is fairly simple and straightforward. Wisdom, and to repeat what was said earlier, being at once a meta-epistemic, meta-ethical and meta-eudaimonic term, provides an immediate and direct conceptual link between information (an epistemic term) and the good life (an ethical-eudaimonic term). As a meta-epistemological term, wisdom as a form of meta-knowledge is capable of providing the necessary reflective knowledge and understanding for evaluating and applying first-order knowledge, which includes both knowledge that (theoretical knowledge) and knowledge how (practical knowledge), in making judgements to reach decisions that concern and impact on different aspects of a person's life (general aspects such as capacities, constraints and circumstances). The ultimate purpose of which is to assist and guide that person, and society more generally collectively, in living a good life for the attainment of well-being. For what would be the point of choosing and leading a life that was not at least in principle capable of providing a good life for the attainment of well-being? Or at least, a life that is overall better with wisdom than without it. A basic pre-supposition of this chapter is that the presence of wisdom is, all things being equal, a better guide for a good life than its absence or its semantic and conceptual opposite, folly.[5] The

[5] Erasmus of course in Praise of Folly makes a case, albeit an ironic and satiric one, that Folly is in fact a better guide than Wisdom for having a good life in this world. On the serious side, Erasmus, however, goes on to suggest in the last section of In Praise of Folly that since only God is capable of wisdom and human beings only capable of folly we should submit ourselves to the will and guidance of God. Being primarily a secular examination of the role of wisdom in a good life, this chapter will not, however, pursue that theological line of inquiry, interesting though it might be. However, the role the Stoic notion of nature plays in a good life for the attainment of eudaimonia, as examined in Chap. 3, is suggestive and of relevance in the neo-Stoic version of that concept presented in those chapters.

notion of wisdom that informs the discussion of this chapter is developed in greater detail in Section "Information, Knowledge and Wisdom" is conceived as forming a continuum comprising degrees of wisdom. At the two extreme conceptual ends of the continuum, we have wisdom (good = 1) and folly (bad = 0). Most people, I speculate, probably fall somewhere in the middle—they are neither wise nor foolish. Yet, another reason why wisdom is relevant to an enquiry concerning the role that information plays in a good life is that some of the issues with regard to the use of digital information qua good life relate not to moral or immoral conduct by an individual towards others but to the prudentially appropriate or inappropriate conduct of the individual in relation to themselves. For example, if an individual posts compromising pictures of themselves on Facebook, Instagram or YouTube, that causes no harm to anyone else but themselves, then the matter is not one of ethics, as such, but one of wisdom or at least prudence[6]—how wise or prudent was the individual to do so, if the outcome of their unwise or foolish behaviour in cyberspace turned out to be harmful to themselves in some way—say, as a result of their unwise or foolish conduct that individual was fired from a job that they enjoyed and for which they showed great potential. In sum, the chapter will seek to show that although related, wisdom is conceptually distinct and conceptually different from both information and knowledge in a crucial way. For wisdom, unlike information and first-order knowledge, provides a person with understanding concerning the techne viou or craftsman ship of living in the sense of knowing how to evaluate and apply relevant information or knowledge for living a good life for the attainment of eudaimonia, and in addition, an appreciation in knowing why such a life constitutes a good life. You will recall that for the Stoics, and generally for the Hellenistic philosophers, philosophy was conceived as a way of life or techne viou. This view of wisdom, as enabling a life worth living, is also similar to the notion of wisdom defended by Sharon Ryan, who in her article "What is Wisdom?" concludes that an accurate answer to that question is that "S is wise if S knows, in general, how to

[6] The term prudence used throughout this chapter refers to the virtue of prudence conceived as an enabling disposition or trait of character that has the tendency of preventing an individual of engaging in conduct that is likely to cause them harm. That is, conduct unbecoming of a virtuous person. Unlike the term instrumental rationality, which is a non-moral term and refers to the non-moral self-interest of an individual, prudence by contrast refers to the virtuous and hence moral self-interest of an individual person.

live well and if S has a general appreciation of the true value of living well" (Ryan 1999). Ryan's views on wisdom will be discussed in more detail in Section "Information, Knowledge and Wisdom", of this chapter.

A Universal Model for Evaluating the Normative Quality of Digital Information

The object of this section of the chapter is to describe and demonstrate in summary a meta-theoretical framework for the normative evaluation of digital information in terms of its inherent epistemological and ethical categories. The primary aim for doing so is to demonstrate in the first instance the conceptual relationship between information and knowledge and the epistemological and ethical commitments to which information understood as a type of knowledge gives rise. As aforementioned, the argument for the inherent normative structure of information as a process of communicative action has been examined and demonstrated at length in a previous publication devoted exclusively to that topic (Spence 2009).[7] In this section of the chapter, the primary aim is to provide a summary of that argument sufficient for establishing an initial theoretical standpoint from which to further motivate and develop the normative relationship between information, knowledge, wisdom and the good life that forms the main topic of discussion in Section "Information, Knowledge and Wisdom". The demonstration of that normative relationship, at once epistemological, ethical and eudemonic,[8] will establish and demonstrate a direct normative link between information and the good life and well-being. Its purpose is to provide a methodological approach in terms of a wisdom theoretic model for the normative evaluation of that relationship and its implications and impact for individuals and societies generally.

The Inherent Normative Structure of Information and Knowledge

In describing the DOIT used for the normative evaluation of information in terms of its inherent epistemological and ethical categories, the chapter

[7] The paper Spence (2009) provides the basis for this part of the chapter.

[8] These three normative terms, the epistemological, ethical and eudaimonic, will sometimes be referred to in this chapter by the collective term, axiological, to be understood as, normatively conferring values.

will employ an epistemological account of semantic information based on a minimal nuclear definition of information (Dretske 1999). Following Luciano Floridi (Floridi 2005), information is defined as "well-formed meaningful data that is truthful" and following Dretske information is defined as "an objective commodity capable of yielding knowledge" and knowledge, in turn, is defined as "information caused belief." The reference to both Floridi's and Dretske's notions of information in this chapter is not intended for making any critical theoretical comparisons between those two accounts of information (which lies beyond the scope of this chapter). It is rather intended for highlighting the one essential element that those accounts have in common, namely, that what is necessary for both information and knowledge is truth. For information without truth is not strictly speaking information but either misinformation (the unintentional dissemination of well-formed and meaningful false data) or disinformation (the intentional dissemination of false "information"). Using the minimal account of information described above, we can now develop an inherent normative account of information, which demonstrates and describes the generic epistemological and ethical commitments that necessarily arise in the dissemination of semantic information, specifically as a process of communication.[9]

Briefly, the argument is as follows (Spence 2009): insofar as information is a type of knowledge (it must be capable of yielding knowledge, one must be able to learn from it) it must comply with the epistemological conditions of knowledge, specifically, that of truth. And insofar as the dissemination of information is based on the justified and rightful expectation among its disseminators and especially its users that such information should meet the minimal condition of truth, then the disseminators of information are com mitted to certain widely recognised and accepted epistemological criteria. Those epistemic criteria will in the main comprise the objectivity as well as the independence, reliability, accuracy and trustworthiness of the sources that generate the information. The epistemology of information, in turn, commits its disseminators to certain ethical principles and values, such as honesty, sincerity, truthfulness, trustworthiness and reliability (also epistemological values), and fairness, including justice, which requires the equal distribution of the informational goods to all citizens. Thus, in terms of its dissemination, as a process of

[9] For a more detailed account of the epistemological and ethical commitments to which information as a process and product of communication gives rise, see Spence (2009).

communication, information has an intrinsic normative structure that commits everyone involved in its creation, production, search, communication, consumption and multiple other uses, to epistemological and ethical norms. These norms being intrinsic to the normative structure of information with regard to all its disseminating modes are rationally unavoidable and thus not merely optional.

Information and Universal Rights

The object of the following associated argument to the one above is to show that in addition of committing its disseminators to unavoidable epistemological and ethical standards by virtue of its own inherent normative structure in terms of truth, information as a process of communication commits its disseminators to respect for peoples' rights to freedom and well-being. This is by virtue of the inherent normative structure of action and specifically informational action, due to its essential features of freedom and well-being (Spence 2006, 2009; Gewirth 1978).[10] Insofar as the communication of information constitutes a type of informational action, information as a process and product of communication must not be disseminated in ways that violate peoples' fundamental rights to freedom and well-being (generic rights), individually or collectively, or undermine their capacity for self-fulfilment. In addition, information must as far as possible be disseminated in ways that secure and promote peoples' generic rights and capacity for self-fulfilment (positive rights) when those rights cannot be secured or promoted by the individuals themselves and can be so secured and promoted at no comparable cost to its disseminators. As we saw in Chap. 3, the concept of self-fulfilment that I am using here, is the neo-Stoic equivalent of the Stoic concept of eudaimonia that I ascribe to Alan Gewirth in that chapter. Due to constrains of space, and because it is beyond the scope and focus of this book on Stoic and neo-Stoic philosophy and its application to technology, I will not attempt to provide a

[10] Alan Gewirth's main thesis in *Reason and Morality* (1978) is that every rational agent, in virtue of engaging in action, is logically committed to accept a supreme moral principle, the Principle of Generic Consistency that commits every agent to respect the rights of freedom and well-being of all other agents including their own. The basis of his thesis is found in his doctrine that action has an inherent normative structure whose necessary features are freedom and well-being, and because of this structure every rational agent, just in virtue of being an agent, is committed to certain necessary prudential and moral constraints and in particular respect for all agents' rights to freedom and well-being.

detailed justification for Alan Gewirth's argument for the principle of generic consistency (PGC) on which his derivation of rights is based.[11] I will, however, offer a brief summary of the rationale of the argument for the PGC by way of a schematic outline of the three major steps of that argument.

The Rights of Agents: Alan Gewirth's Argument for the PGC

Gewirth's main thesis is that every rational agent, in virtue of engaging in action, is logically committed to accept a supreme moral principle, the PGC. The basis of his thesis is found in his doctrine that action has a normative structure, and because of this structure every rational agent, just in virtue of being an agent, is committed to certain necessary prudential and moral constraints. Gewirth undertakes to prove his claim that every agent, qua agent, is com mitted to certain prudential and moral constraints in virtue of the normative structure of action in three main stages (Gewirth 1996). First, he undertakes to show that by virtue of engaging in voluntary and purposive action, every agent makes certain implicitly evaluative judgements about the goodness of his purposes, and hence about the necessary goodness of his freedom and well-being, which are the necessary conditions for the fulfilment of his purposes (Gewirth 1998). Second, he undertakes to show that by virtue of the necessary goodness which an agent attaches to his freedom and well-being, the agent implicitly claims that he has rights to these. Third, Gewirth undertakes to show that every agent must claim these rights in virtue of the sufficient reason that he is a prospective purposive agent (PPA) who has purposes he wants to fulfil. Furthermore, every agent must accept that, since he has rights to his freedom and well-being for the sufficient reason that he is a PPA, he is logically committed, on pain of self-contradiction, to also accept the rational generalisation that all PPAs have rights to freedom and well-being. At this third stage of the argument, these rights, being not only self-regarding but also other regarding, are now moral rights. The conclusion of Gewirth's argument for the PGC is in fact a generalised statement for the PGC, namely, that all PPAs have universal rights to their freedom and well-being (Gewirth 1978). Applying the PGC to information, we can now make the further

[11] A full and detailed defence of the argument for the PGC against all the major objections raised against it by various philosophers can be found in Spence (2006, Chapters 1–3), Deryck Beyleveld (1991), and Alan Gewirth (1978).

argument that information generally, including both analogue and digital information, must not be disseminated in ways that violate informational agents' rights to freedom (FR) and well-being (WB), individually or collectively (negative rights). Moreover, information must as far as possible be disseminated in ways that secure and promote the informational agents' rights to FR and WB (positive rights). Conceived as the Fourth Estate, this places significant and important responsibility on the disseminators of information and in particular the media, especially journalists, both offline and online, as well as the social media. In conclusion of this section, the DOIT demonstrates the dual normative structure of informational action, to which all informational agents, including the media (the Legacy Media of the 4th Estate, the Digital Media of the 5th Estate (Elliott and Spence 2018) and the Big Tech Network Media of the 6th Estate such as Facebook and Google, a topic we shall examine in Chap. 6, are committed by universal necessity. Information generally can be epistemologically and ethically evaluated internally by reference to its inherent normative structure. Insofar as the ethical values to which the inherent normative structure of information gives rise requires that the informational agents' rights to FR and WB should be respected, secured and promoted, those values are also mandated by the PGC and thus information can also be externally evaluated by reference to the PGC. Expressive Information can also be evaluated either internally or externally or both, in this way. For example, identity theft on the Internet is morally wrong both because it is untruthful (internal evaluation) and because it can cause harm (external evaluation).

Information, Knowledge and Wisdom

This is the focal and main section of the chapter whose aim is the examination and evaluation of the relationship between the concepts of information, knowledge and wisdom.[12] Approximately and pending further

[12] Surprisingly, very little has been written on the relationship between information, knowledge, and wisdom in the philosophical literature, specifically with regard to information ethics. The main sources I will refer to in this part of the paper, but not exclusively, are Maxwell (2007), Tiberius (2008), Vitec and Jackson (2008), Varelius (2004), Kvanvig (2003), Finnis (1980, 1983), Ryan (1999, 2014), Kekes (1995); and generally the writings of Plato (Apology), Aristotle (Nichomachean Ethics) and the Hellenistic Philosophers (Epicureans, Sceptics, and Stoics)—for a discussion of those see Spence (2006, Chapter 10). See also, Vallor (2016). Technology and the Virtues: A Philosophical Guide to a Future Worth. Wanting, Oxford: Oxford University Press.

discussion below, wisdom in this chapter is conceived as a type of meta-knowledge that is used in the evaluation and application of information and knowledge to make right judgements in reaching appropriate decisions that are of value and good for us in our lives personally (prudentially and eudaimonically good) and that are of value and good for others in their lives (ethically good) for the ultimate attainment of a good life resulting in eudaimonia. The overall objective of this line of enquiry is to further determine what the relationship between information and the good life is and to what degree, if any, information, in its dissemination and communication, through its related enabling ICTs and AI technologies, can in principle contribute to a good life for the attainment of eudaimonia. In this chapter, we examine more closely how information can be directly related to the notion of a good life via the concept of wisdom: if wisdom is a primary and essential condition for an individual in (a) determining what a good life is or ought to be (meta-knowledge-that and meta-knowledge-why) and (b) a primary and essential condition in providing us with guidance and direction, both as individuals and societies generally, of how to live such good lives and (c) moreover, wisdom, as a reflective meta-virtue is an enabling disposition of character that practically enables us to live such good lives for the attainment of eudaimonia (meta-knowledge how), to what extent and in what ways, if any, does information contribute to wisdom and by extension to the good life? The chapter posits that one direct way of evaluating the value of information and its relation to a good life generally (its overall axiological goodness) is by determining the degree to which it contributes or is capable of contributing to the attainment of a good life epistemologically (its capacity to yield knowledge), ethically (its ability to contribute to the moral good of others both negatively by causing no unjustified harm to others, and positively by causing positive good for others) and eudaimonically (its capacity to con tribute to both the conception and the attainment of a good life). The chapter will show that to achieve that theoretical objective the notion of wisdom is essentially required. Regarding the concept of a good life, we can say quite reasonably that a good life generally is one that is at least minimally capable of enabling a person to attain self-fulfilment, well-being, happiness or eudaimonia. As aforementioned, I will for methodological convenience use the term eudaimonia[13] to include and refer to all those concepts collectively while maintaining the original meaning for that term as intended by the ancient Greek philosophers, including Plato, Aristotle and the Hellenistic philosophers and in particular the Stoics, whose philosophy was the topic examined in Chap. 3. Although those philosophers might have explained the notion and attainment of eudaimonia in different ways, they all at least agreed that the virtues were essential (and for the Stoics also sufficient) for the attainment of a eudaimonic or a flourishing life and moreover, the virtues were constitutive of such a life. For insofar as eudaimonia is the ultimate object in life as Aristotle claimed, it is difficult to conceive a life that was not at least capable of leading to the attainment of eudaimonia, as good—what would it be good for if it were incapable of at least in principle enabling one to realise one's ultimate objective in life? Analysing information through the application of the concept of meta knowledge (knowledge-that, knowledge-why and knowledge-how) of what is good or evil for us and others—how it contributes or is capable of contributing to a good life for us and others for the attainment of eudaimonia—is what the chapter will initially postulate as Wisdom. In sum, if a good life should at least be capable of leading to self-fulfilment or eudaimonia (otherwise what is it good for?), especially, self-fulfilment as capacity fulfilment (making the best of oneself as a human being[14] then wisdom (understood as a type of meta-knowledge, the acquisition of which enables one to create, communicate and use information so as to render oneself and others, whenever possible, capable of achieving self-fulfilment and eudaimonia) is a necessary condition for a good life. An important qualification to the claim made in this chapter that wisdom is a necessary condition for a good life is that such a life is conceived eudaimonically. For the notion of wisdom developed in this chapter and applied in evaluating the axiological

[13] For the purposes of this chapter, I will use the notions of eudaimonia and eudaimonic pluralistically as being potentially compatible with various different theories of the good life, including hedonistic, desire-satisfaction and objective-list theories, among others. Simply put, the notion of a good life used in this chapter is a good life that is in principle capable of leading to the attainment of eudaimonia. As such, any theory of a good life capable of leading to eudaimonia can at least in theory and upon further demonstration be considered a eudaimonic life. My own theoretical preference is a neo-Stoic eudaimonic life that includes the virtues but that need not exclude other theories capable of also leading to the attainment of eudaimonia. See, for example, Vallor (2016). Technology and the Virtues.

[14] See Chap. 3 of this book that demonstrates how Gewirth's notion of self-fulfilment, as capacity fulfilment, is conceptually similar to the Stoic notion of eudaimonia.

goodness of information is itself a eudaimonic conception of wisdom. However, such a eudaimonic notion of wisdom is not unlike our common sense and pre-theoretical understanding of wisdom, namely, an overarching reflective capacity the possession of which allows one to lead a good life and, moreover, enables one to guide others in leading fulfilling and good lives. This eudaimonic notion of wisdom is akin to the notions of wisdom defended by philosophers such as Plato, Aristotle, the Epicureans, and as we saw in Chap. 3, by the Stoics and neo-Stoics, who although postulated and defended somewhat different notions of the good life, can nevertheless collectively be thought of as offering eudaimonic accounts of the good life. It can be said that a common denominator for all the cited eudaimonic accounts of wisdom is their subordination of the concepts of pleasure and desire to that of virtue. The essential link between pleasure and desire on the one hand and virtue on the other might be weaker in the case of the Epicureans and stronger in the case of the Stoics but whatever the strength of that relationship might be, the link between pleasure, desire and virtue is an essential characteristic of traditional eudaimonic conceptions of a good life and also eudaimonic conceptions of wisdom. Moreover, as mentioned above, another essential characteristic of a eudaimonic conception of a good life is that such a life is capable of resulting in the attainment of eudaimonia. As such, the term eudaimonic understood in this very general sense is compatible with a number of other theories of the good life which can be shown to be capable of leading to the attainment of eudaimonia. This is in keeping with the Eudaimonist axiom, the view that "happiness is desired by all human beings as the ultimate end or telos of all rational action."[15] Importantly, the relationship between wisdom and a good life proposed in this chapter under a eudaimonic conception of a good life is reflexive. For wisdom guides one to the choice of a eudaimonic conception of a good life and the pursuit of such a life, and a eudaimonic conception of a good life, in turn, guides and motivates one to the acquisition of wisdom as an enabling disposition, in the form of an overarching reflective virtue, which is necessary for the attainment of a eudaimonic life. This should not surprise us. For although wisdom acts initially instrumentally, as a necessary enabling virtuous disposition for the attainment of a eudaimonic life, once attained, a eudaimonic life becomes inseparable from the state of wisdom that enabled its attainment. This

[15] Brink attributes the Eudaemonist axiom to Gregory Vlastos in Vlastos (1991) Socrates: Ironist and Moral Philosopher. Ithaca: Cornell University Press, p. 203.

reflexivity between eudaimonia and wisdom allows us then to say that a wise person is generally a eudaimonic person, and a eudaimonic person is generally a wise person. However, I don't wish to exclude the logical possibility that one could be wise but unhappy although pragmatically, given our common understanding of wisdom, that would be an odd thing to say and in practice I think, an unusual occurrence. For as shall see in Chap. 3, the Stoics claimed that the virtues, including practical wisdom or prudence, was not only necessary for a eudaimonic life but also sufficient. As we also saw in Chap. 3, the neo-Stoic philosophy I am attributing to Alan Gewirth through his theory of self-fulfilment, also supports the sufficiency of virtue for a eudaimonic life.

What Is Wisdom?

Having examined in some detail what information and knowledge are and what the relationship is that exists between them (by way of an examination of the essential property that characterises both, namely, the property of truth) it is now time to turn our attention to the notion of wisdom so as to explore further the conceptual relationship that holds between information, wisdom and a good life. According to Nicholas Maxwell (2007), The central task of inquiry is to devote reason to the enhancement of wisdom—wisdom being understood here as the desire, the active endeavour and the capacity to discover and achieve what is desirable and of value in life, both for oneself and for others. Wisdom includes knowledge and understanding but goes beyond them in also including: the desire and active striving for what is of value, the ability to see what is of value, actually and potentially, in the circumstances of life, the ability to experience value, the capacity to help solve those problems of living that arise in connection with attempts to realise what is of value, the capacity to use and develop knowledge, technology and understanding as needed for the realisation of value. Wisdom, like knowledge, can be conceived of, not only in personal terms, but also in institutional or social terms. We can thus interpret the philosophy of wisdom as asserting: the basic task of rational inquiry is to help us develop wiser ways of living, wiser institutions, customs and social relations, a wiser world. What is of interest in Maxwell's quoted passage for our present purposes is the relationship he draws between the concepts of reason, knowledge, understanding and the desire, capacity and active endeavour for the achievement (or as in my case attainment) of what is of value in life, for oneself and others. With the exception of understanding, for which I will have more to say in what follows, the other concepts to which Maxwell, draws attention seems to anticipate and reflect both explicitly and implicitly, the concepts included in my own normative analysis of information and knowledge, in terms of their epistemological, ethical and eudaimonic dimensions. The basis of that analysis is the meta-theoretical framework comprising the DOIT and, as we shall see, the DOIT-Wisdom model.

Four Theories of Wisdom

In an article in the Stanford Encyclopedia of Philosophy, Sharon Ryan (2014) identifies at least four different theories of wisdom: (A) Wisdom as Epistemic Humility, which she attributes to Socrates in Plato's Apology (Plato 1978). There Socrates expresses puzzlement concerning the oracle of Delphi's pronouncement that he is the "wisest of men" and declares that his knowledge extends only as far as his knowledge of his own ignorance, the prototype case of epistemic humility (B) Wisdom as Epistemic Accuracy, for which Ryan provides two versions to the effect that (B1) S is wise iff for all p, (S believes S knows p iff S knows p) (EA1) or the weaker version (B2) S is wise iff for all p, (S believes S knows p iff S' belief in p is highly justified.) (EA2); (C) Wisdom as Knowledge: the view that knowledge is at least a necessary condition of wisdom. Ryan identifies several philosophers who hold some version of (C) including Aristotle (Nichomachean Ethics Artistole 1941), Descartes (Principles of Philosophy), Richard Garrett (1996), John Kekes (1983), Lehrer and Smith (1996), Nicholas Maxwell (1984), Robert Nozick (1989), Plato (The Republic) (Plato 1978) and Sharon Ryan (1996, 1999). According to Ryan, all these philosophers "have theories of wisdom that require a wise person to have some knowledge of some sort" and what's more, "all these views maintain that wise people know what is important" (Ryan 2014). Overall, these theories differ according to Ryan "over what it is that the wise person must know and whether there is any action that is required for wisdom" (Ryan 2014). To further differentiate different notions of wisdom that fall under the broad category of "wisdom as knowledge," Ryan refers to Aristotle who held two main theories of wisdom, Sophia, or theoretical wisdom, and Phronesis, or practical wisdom. Theoretical wisdom according to Aristotle is "scientific knowledge, combined with intuitive reason, of the things that are highest by nature" (Ryan 2014 quoting Aristotle in Nichomachean Ethics, VI, 1141b). On that basis, Ryan interprets Aristotle's notion of theoretical wisdom as the following view: (C1) Wisdom as extensive factual knowledge (WFK) which in effect amounts to "S is wise iff S has extensive factual knowledge about science, history, philosophy, literature, music, etc." (Ryan 2014). Ryan finds this notion of wisdom implausible for as she correctly observes "some of the most knowledgeable people are not wise" (Ryan 2014). Ryan maintains that Aristotle's notion of phronesis or practical wisdom is a more reasonable theory to hold. According to Aristotle, Now it is thought to be the mark of a man of practical wisdom to be able to deliberate well about what is good and expedient for himself, not in some particular respect, for example, about what sorts of things conduce to health or strength, but about what sorts of things conduce to the good life in general. (Nichomachean Ethics, VI, 1140a01140b) Ryan concludes that "for Aristotle, practical wisdom requires knowing, in general, how to live well." She goes on to say that although many philosophers are in agreement with Aristotle on this point they would not agree with Aristotle "that theoretical wisdom is one kind of wisdom and practical wisdom another. Wisdom, in general," Ryan concludes, "requires practical wisdom" (2014). In support of this general notion of practical wisdom, Ryan cites Robert Nozick who claims that "Wisdom is what you need to understand in order to live well and cope with the central problems and avoid the dangers in the predicaments human beings find themselves in" (Nozick 1989). She also cites John Kekes, whose view is that "what a wise man knows, therefore, is how to construct a pattern that, given the human situation, is likely to lead to a good life" (Kekes 1983). Ryan defines this type of general practical wisdom as (C2) Wisdom as Knowing How to Live Well (KLW): "S is wise iff S knows how to live well," which according to Ryan captures on the whole Aristotle's concept of practical wisdom as well as the views held by Nozick, Plato, Garrett, Kekes and Ryan (2014). The final theory of wisdom that Ryan considers is Wisdom as Knowledge and Action, which she specifically defines as Wisdom as Knowing How To, and Succeeding at, Living Well (KLS), which in effect amounts to: S is wise iff S knows how to live well, and (ii) S is successful at living well (2014). According to Ryan, the "idea of the success condition [condition (ii) in KLS] is that one puts one's knowledge into practice." She goes on to attribute a view broadly along the lines of (KLS) to Aristotle (his notion of practical wisdom), as well as to Kekes and Nozick. Ryan herself rejects this theory based on criticisms she raises in (Ryan 1999) but her main criticism with which I concur is that (KLS) seems to leave out the factual knowledge required by the theory of WFK. In agreement with Ryan, I also claim that some factual knowledge of the world (but not necessarily extensive) adequate for enabling a person to make their way in the world and have a good life, is necessary for wisdom. This consideration introduces an important distinction when enquiring into the conceptual connection between knowledge and wisdom: the distinction between knowledge for wisdom and knowledge as wisdom. Although related, the two are quite different and their difference highlights an important and crucial distinction between knowledge and wisdom. No doubt some general knowledge about the world acquired on the basis of reliable and veridical information that causes it and sustains it (Dretske 1999) is necessary for wisdom. This is in keeping with the notion of Socratic ignorance, roughly understood here as having knowledge of one's ignorance (being aware of one's ignorance and humbly acknowledging one's lack of knowledge). For Socratic ignorance prompts and motivates one to acquire the knowledge of which one is ignorant (knowledge understood here as some minimal general knowledge about some basic aspects of the world, e.g., history, geography, science, mathematics, literature, art, etc.). Socratic ignorance as a special type of knowledge accords with Ryan's theory of wisdom as epistemic humility, which was examined earlier. By contrast, those who claim to know what they lack knowledge of, are not in a position to be motivated to acquire the knowledge they lack; and moreover, the knowledge that is at least in a minimal and general sense partly necessary for the acquisition of wisdom and by extension, the attainment of a good life and eudaimonia. Thus, at a minimum, and bracketing the possibility that "holy fools" though totally ignorant of facts about the world are nevertheless in some sense "wise," some minimal and general knowledge about the world is instrumentally and prudentially necessary for the acquisition of wisdom. At least at a minimum, an attitude of Socratic ignorance[16] might be necessary for the acquisition of wisdom. For the Socratic elenchus can be applied as a method for acquiring the knowledge one lacks, through first recognising and acknowledging one's ignorance, and then being motivated to gradually acquire the knowledge of which one is ignorant, through critical enquiry and further investigation. According to John Kekes, "the elen-

[16]For a detailed discussion of the Socratic elenchus and Socratic wisdom, see Hugh H. Benson (2000). Socratic Wisdom: The Model of Knowledge in Plato's Early Dialogues. Oxford University Press.

chus enables its practitioners to progress from a special kind of ignorance – foolishness – to a special kind of knowledge – moral wisdom" (Kekes 1995). We can therefore say that the acquisition of such general minimal knowledge about the world or an attitude of Socratic ignorance when we lack such knowledge is instrumental to the acquisition of wisdom because it provides at least part of the necessary means, that is, the capacity for the acquisition of wisdom. Moreover, the acquisition of such minimal and general knowledge of the world or in its absence, an adoption of an attitude of Socratic ignorance, is prudential to the acquisition of wisdom. Insofar as we consider the acquisition of wisdom desirable, valuable and essential for the attainment of a good life, we should (normatively) inculcate in ourselves the virtue of learning: the desire and active pursuit of the acquisition of at least a minimal and general knowledge about the world. Hence, some minimal and general knowledge about the world is necessary for wisdom. The Stoics also claimed that some general knowledge of the world and more generally knowledge of nature was also required for wisdom. In the Meditations, for example, Marcus Aurelius speaks "of an understanding which comprehends the inmost being of each thing, its place in the world-order, the term if its natural existence, the structure of its composition" (Aurelius 1964, Book 10:9, 155). In addition, along with Aristotle, I also wish to claim that being a good per son is a necessary condition, for being wise. For Aristotle claims, "Therefore it is evident that it is impossible to be practically wise without being good" (Nichomachean Ethics, VI 1144a). Contrary to Aristotle, however, and in accordance with the Stoics and my Gewirthian reconstruction of a neo-Stoic position I argued for in Chap. 3, I wish to claim that being a good person, that is, a person in possession of a virtuous character, is also sufficient for being wise, in successfully applying wisdom in living well, at least for the aim, if not the actual successful attainment of eudaimonia, a prize reserved only for the Stoic sage, for example, someone like Socrates. John Kekes seems to also have the view that a virtuous character is an essential characteristic of the wise person. In the opening sentence of his book Moral Wisdom and Good Lives (1995, ix), Kekes tells us that "moral wisdom is a virtue – the virtue of reflection." A more detailed characterisation of moral wisdom by Kekes (1995) is that moral wisdom is the capacity [a psychological capacity] to judge rightly what should be done in particular situations to make life better…. Because this human psychological capacity, once developed, is likely to be lasting and important, it can be identified as a character trait …. We can say, therefore, that people have moral wisdom if they regularly and predictably act wisely in the appropriate situations and if so acting is an enduring pattern in their lives .... Whether an action is morally wise depends also on what the agents bring to the judgements they make, such as their particular conception of what would make life better. According to Kekes, moral wisdom is a second order virtue whose primary concern, [Is] the development of our character [emphasis added] in a desirable direction by strengthening or weakening some of our dispositions. First-order virtues guide our actions in view of what we think of a good life; second-order virtues guide our actions with a view of developing the kind of character that reflects a reasonable conception of a good life (1995). Kekes' claims to moral wisdom above are also in keeping with the importance the Stoics and neo-Stoics attribute to character as an essential aspect of being a good and wise person.

The Psychology of Wisdom: The Berlin Wisdom Paradigm

Finally, it is worth mentioning a psychological theory of wisdom that runs parallel to the philosophical theories of wisdom discussed above. This is the theory of wisdom postulated by Paul Baltes and his research associates from the *Max Planck Institute for Human Development* in Berlin. The theory known as the *Berlin Wisdom Paradigm* views wisdom as a kind of expertise in the matters of human life (Baltes 2004; Baltes et al. 2002; Baltes and Smith 1990 as referred to by Banicki 2009). In a chapter from A Handbook of Wisdom: Psychological Perspectives (Robert J. Sternberg and Jennifer Jordan 2005) , Ute Kunzmann and Paul B. Baltes define wisdom in accordance with the Berlin Wisdom Paradigm, as a "highly valued and outstanding expertise in dealing with fundamental, that is, existential, problems related to the meaning and conduct of life" (2005). Kunzmann and Baltes go on to elaborate that the "focus of their theoretical work has been to define wisdom as an expert system in human thought and behavior that coordinates knowledge and virtue, mind and character" [emphasis added] (2005), and that thus defined, wisdom for them "reflects both components of wisdom: intellect and character" (2005). According to them, "wisdom differs from other human strengths in that it involves an orchestration of mind and virtue, intellect and character" (2005). It is not too much of a jump to reasonably interpret the psychological notion of the Berlin Wisdom Paradigm as expressed above by Kunzmann and Baltes as parallel and broadly in keeping with the general view of practical wisdom, as knowledge of how to live well and the successful application of that knowledge in living well. However, with the additional proviso that the knowledge in question must also include some knowledge of facts concerning the world, which is Ryan's notion of wisdom (KLS) we examined above. Such knowledge does involve an "orchestration," as in the case of Kunzmann and Baltes' psychological notion of wisdom, of "intellect" or "cognition" (the epistemological features) and virtue and character (the ethical and eudaimonic features) in my tripartite model of wisdom, conceived as meta knowledge and meta-virtue, which, in general, is in keeping with the notion of intelligent virtue developed by Julia Annas in her book by that title (2011). Moreover, the theories of wisdom we examined above, as well as that of the Berlin Wisdom Paradigm, are generally in keeping with Stoic and neo-Stoic philosophy we will examine in Chap. 3.

Wisdom and the Evaluation of Digital Information in Relation to a Good Life

The outcome of the extended argument concerning the nature of wisdom, what wisdom is and its relationship to information and knowledge in Information, Knowledge and Wisdom, is that wisdom is a special type of meta-knowledge and meta-virtue. Insofar as the ultimate purpose of a good life is the attainment of eudaimonia then wisdom, which informs the conception of a good life and directs its active pursuit for the attainment of eudaimonia, is an essential condition for both the conception and the attainment of a good life. As the essential condition for both the conception and guided active pursuit and successful achievement of the good life, wisdom is therefore established as the essential conceptual connection between information and the good life and in particular information that is minimally necessary for acquiring knowledge of the things we need, to have a good life. Some general knowledge about the world acquired on the basis of reliable and veridical information that causes it and sustains it (Dretske 1999), is therefore necessary for wisdom. This, in turn, allows us to determine some of the generic implications and ramifications of information for the conception of a good life, in particular, a eudaimonic conception of a good life. However, as Kekes points out, "the eudaimonic conception of a good life is not to be understood as the endorsement of a particular form of life. It is rather a regulative ideal that specifies some general conditions to which all good lives must conform" [emphasis added] (Kekes 1995). As such, the eudaimonic account of a good life canvassed in this chapter is broadly speaking pluralistic as it is in principle compatible with other different conceptions of a good life that meet the same necessary general conditions to which any notion of a good life must conform. For example, insofar as hedonistic, desire-satisfaction and objective list theories of the good life meet the minimal conditions for both specifying what a good life is, as well as providing the enabling conditions for its practical realisation, then they too can be aligned broadly to the notion of wisdom developed in this chapter. Kekes' claim cited above is insightful and very much in keeping with the eudaimonic conception of a good life proposed in this chapter on the basis of the DOIT-Wisdom. For the DOIT-Wisdom model (2011) is intended only as a meta-theoretical regulative ideal that specify some general conditions to which all good lives (and in particular informational lives) must conform regardless of the particular contexts and contingencies of those lives. Specifically, with regard to the creation and dissemination of information, central to those general conditions to which all good lives are bound are (a) the epistemological and ethical obligations that emanate directly from the inherent normative structure of information; (b) the universal rights to freedom and well-being to which all agents are entitled and which arise naturally from the inherent normative structure of informational action and (c), the virtues of character and associated moral sentiments and values that are prudentially desirable and required as enabling general motivational dispositions for the pursuit of a good life for the ultimate attainment of eudaimonia. In sum, we can say that those general meta-conditions are encapsulated within the combined models put forth in this chapter, namely, those of DOIT and Wisdom to which I refer to collectively in this chapter as the DOIT-Wisdom model (Spence 2011). That model moreover seems adequate for the normative evaluation of an informationally good life for the attainment of well-being, as it combines a tripartite structure comprising the three essential normative principles for the evaluation of digital information, namely, the epistemic, the ethical, including the virtues as enabling dispositions, and the eudaimonic. Kekes correctly claims that according to a eudaimonistic conception of a good life, Primary values [values that concern uniform and universal human goods and needs] may be thought of as establishing the moral limits and secondary values [values that vary across individuals in accordance with differences in cultural traditions, conceptions of a good life, and individual contingencies and circumstances] as establishing the moral possibilities that define good lives (1995) ... the former define a grid [emphasis added] within which human beings must endeavour to make a good life for ourselves, while the latter provide the ways in which individuals fill in the grid (1995). It has been the overall objective of this chapter to provide such a conceptual grid; namely, the DOIT-Wisdom model. Having discussed meta-theoretically the epistemological, ethical and eudaimonic implications of information as a process of communication for individuals and society generally through the notion of wisdom in Section "Information, Knowledge and Wisdom", I will in this final section of the chapter apply the DOIT-Wisdom model to identify some of the specific implications and ramifications that the production, dissemination and communication of digital information might have for the good lives of individuals and society generally. Before doing so, it would be useful to provide a schematic representation of the triadic structure of the DOIT-Wisdom model: The DOIT-Wisdom model (Spence 2009, 2011) enables the normative evaluation of technology in relation to well-being based on the concept of wisdom. The argument presented in this chapter demonstrates that the notion of wisdom understood as triadic notion comprising at once a meta-epistemic, meta-ethical and a meta-eudaimonic normative categories, provides the essential conceptual link between technology (specifically, information and communication technologies and AI technologies) and well-being.

Why Wisdom Is of Relevance to the Evaluation of Digital Information

The model DOIT-Wisdom presented in this chapter for the normative evaluation of information and its relation to the good life is a model that applies to all information whether digital or analogue regardless of whether such information is generated within the legacy media or the digital and social media. However, the exponential growth of digital information on the Internet, through technology networks, such as those Facebook, Google, Microsoft, Apple and Amazon, collectively referred to in this book as Big Tech or the Big Five, and its implications for the good life of individuals and society generally, merits special attention for at least three theoretical reasons. There is, to begin with, a metaphysical or ontological reason: the digitalisation of information has extended the natural and technological boundaries of our lives to such an extent that as Luciano

Floridi (Floridi 1999) correctly observes we now live in the infosphere (2002) and not just the biosphere. We have essentially become and are increasingly becoming informational beings. As such, billions of people around the world spend a great deal of their lives, thinking, communicating, working, buying, selling, banking, playing, socialising and generally, acting informationally. All these numerous and varied informational activities in the infosphere, and specifically the digital infosphere, are unavoidably related to the quality and goodness of our lives, both as individuals and collectively as societies. Hence, if wisdom is the conceptual link that normatively links information to a good life and well-being, as argued in this chapter, clearly then wisdom has a direct and primary relevant role to play in the normative evaluation of digital information and its relationship to a good life and well-being in the infosphere. As shown in Chap. 3, wisdom as a core principle of Stoic philosophy applies to the biosphere and analogously it applies equally to the infosphere. Second, there is a technological reason. The digitisation of information allows us countless means and ends for instantly creating and disseminating information around the globe through the Internet and through technological devices such as smartphones, for example. This unprecedented rapid and extensive dissemination of digital information can and does affect the quality of our lives both positively but also negatively. People can both benefit themselves and others through the use of digital information, but they can also harm themselves and others through the unreflective misuse and abuse of information, though, for example, fake news, deep fakes and other forms of misinformation and disinformation. The age of abundant information is paradoxically marked by a deficit of wisdom. It seems that, the more information we have, the less wise we are in managing and controlling it for our individual and collective well-being. The problem is that there is too much information and not enough time to absorb it, understand its implications and judge the best way to use it for our individual and common good. The glut of information has created gluttony for information, which can lead us to behave unwisely and sometimes foolishly. Examples of such unwise and foolish online behaviour abound. Take for example the Australian treble Olympic gold medal winner Stephanie Rice,[17] who lost a lucrative sponsorship with Jaguar as a result of a thought less tweet about the South

[17] Stephanie Rice. https://www.smh.com.au/sport/swimming/i-want-you-to-know-how-sorry-i-am-tearful-rice-20100908-150s3.html. Accessed: April 25, 2021

African rugby team; the Canberra Raiders star Joel Monaghan[18] who was photographed performing an act of simulated bestiality with a dog, which was later published on the Internet and forced his resignation; Catherine Daveny,[19] a journalist with the Age who was fired for a series of unsavoury but mostly silly tweets about various TV personalities and celebrities; a journalist photographer from Ireland who was sacked after making comments on her Facebook page about the young woman Michaela Harte, who had been murdered on her honeymoon in Mauritius; American journalist Nir Rosen,[20] who had to resign after making offensive tweets about Lara Logan, the CBS journalist who was sexually attacked in Egypt. More commonly still, young people undermine their privacy and chances of future employment by placing compromising photos of themselves and friends on Facebook and engaging in cyber-bullying that has allegedly driven some young people to suicide. The problem is widespread and global. There are countless daily informational acts of unwise and self-defeating behaviour, which, but for the all-seeing eye of the omnipresent Internet, would pass unnoticed as matters of no consequence. An important truth the WikiLeaks controversy has revealed is that the Internet and the digital informational environment affords no one the certainty or comfort of privacy. The Internet is, by its very nature, a boundary-free public space not well-suited to private conversations and secrets and, what's more to the point, informational indiscretions. The digitisation of information has fundamentally changed not only the way we disseminate information but the way we live. The Internet, just like the natural environment, cannot be neatly constrained and controlled by any one group of individuals, conglomerates or nations. And like the natural environment, the Internet is capricious and unpredictable and only favours the uninhibited free dissemination of information by anyone, anytime and anywhere. If we cannot control or manage the flow of information on the Internet, the next best thing is to control our own online informational behaviour. That is within our control and in keeping with the Stoic and neo-Stoic philosophy examined in Chap. 3, a central principle of which is wisdom. We have to learn

[18] Joel Monaghan. https://www.smh.com.au/sport/nrl/joel-monaghan-in-tears-after-quitting-the-raiders-20101109-17lff.html. Accessed: April 25, 2021

[19] Catherine Deveny. https://www.smh.com.au/entertainment/tv-and-radio/deveny-dropped-as-columnist-for-the-age-20100504-u6si.html. Accessed: April 25, 2021.

[20] Nir Rosen. https://www.news.com.au/world/journalist-nir-rosenquits-after-insensitive-and-offensive-lara-logan-tweets/news-story/356d46509da0ee10a5d705c226087ee6. Accessed: April 25, 2021.

how to use and disseminate information wisely, in a manner that protects and promotes our individual and collective well-being. Wisdom, as a higher type of knowledge (knowing how to understand and use information upon reflection and with good judgement for the benefit of oneself and society), provides practical "know-how" for applying information to improve our lives and that of others. It is also a reflective virtue in the form of practical prudence, which can teach us how to create and use information to live good and meaningful lives in the infosphere—lives that are capable of leading to self-fulfilment, eudemonia and happiness for us and others. What wisdom requires is that we learn the husbandry of information. How to reflect upon it, how to understand it, how to control it so it does not control us, how to judge its implications so we can foresee its consequences, whether they are good or bad, and how to use it in ways that enhance our well-being and promote and protect our rights to freedom, privacy and respect. In the age of information, it seems that we would be better off with more wisdom and a little less information. Switching over our iPhone and iPad to Plato and Epictetus, sometimes, may be a good start.[21] The uncritical and sometimes thoughtless dissemination of a lot of trivial and personal information of oneself and others on social networks and plat forms such as Facebook, YouTube and Instagram, as well as through SMS and Twitter, as the above examples illustrate, might be not only not conducive to wisdom and a good life overall, but detrimental to it. For they appear to encourage, sometimes at least, unreflective, foolish and reckless behaviour with no apparent compensating axiological or eudemonic value for oneself, one's "friends" and others. Given the pitfalls of creating and accessing unreflectively information about oneself and others, without any compensating realisation of value in relation to one's conception of a good life, a wise person would thus be best served in exercising caution when using digital information. Such overall reflective caution as we saw in Chap. 4, is also in keeping with Stoic philosophy's essential characteristic, namely, control; that is to say, the ability to narrow down the areas in one's life over which one has little or no control so as to enlarge the areas of our lives over which we have control so that we can order them in conformity with our conception of a good life for the attainment of eudaimonia or well-being. This is a process of self-knowledge, a process of reflection involving judgements "whose

[21] Elliott and Spence, Ethics for a Digital Era, Chapter 10; and Spence, Wisdom and Well-being in a Technological Age.

aim is to make our character less fortuitous and more deliberate" (Kekes 1995). When we allow ourselves to unreflectively misuse digital information of ourselves and others on the Internet and through the use of smartphones and other digital devices, we make ourselves more "fortuitous" and less "deliberate." In so doing, we become hostages to fortune, since we no longer have any control over that information, which now forever floats beyond our control in cyberspace. Surely such unreflective conduct is not only not wise but can in extreme circumstances be foolish and ultimately self-defeating if, upon careful reflection, it undermines our considered conception of a good life and that of its eudaimonic realisation. A third reason why wisdom is of primary relevance to an enquiry concerning the role that digital information plays in a good life is that some of the issues with regard to the use of digital information qua good life relate not to moral or immoral conduct by individuals towards others as such, but to the prudentially appropriate or inappropriate conduct of the individuals in relation to themselves, or more precisely what is good for them—good for their lives and well-being. In short, such conduct might only constitute unwise or foolish behaviour and not necessarily unethical conduct. For example, if an individual posts compromising pictures of themselves on Facebook that causes no harm to anyone else but themselves, then the matter is not one of ethics as such, but one of wisdom or at least prudence[22]—how wise or prudent was the individual to do so, if the outcome of their unwise behaviour in public cyberspace turned out to be harmful to themselves in some unintended way. Say, as a result of their unwise but not necessarily unethical conduct that individual was fired from a job they enjoyed and relied upon for their self-fulfilment. Digital media, including our multiple informational interactions on the Internet and through technology networks, such as Facebook, Instagram, Google, YouTube and Twitter, have made it possible for us to access and use extraordinarily and unprecedented large quantities and varieties of information and data. This is indeed the age of information and Big Data. However, what seems to follow from our characterisation of wisdom above is that the uncritical access and use of so much information without the appropriate reflection, judgement and understanding, might not be conducive to wisdom and consequently might not be conducive to a good life for individuals and society. Thus, the uncritical accumulation and use of more information is not necessarily conducive to more wisdom (because not essential to it) and

[22] For the term prudence see Note 6.

hence not more conducive to a good life. On the contrary, sometimes we might be better off with less information rather than more, especially if the former is directed by wisdom (less information) and the latter is not (more information). Even in the case of critically accessing and using contextualised information, such information might also not be conducive to wisdom and to a good life, if that information is merely used instrumentally without a clear understanding of the ends which that information is intended to achieve or a clear understanding of the value of those ends. Recall that wisdom not only directs the means but also the ends of our actions. Thus, the accessing and use of a lot of contextualised information on our smartphones, for example, without a clear understanding of the value of the ends for which that information is to be used, might at best be neutral with regard to the goodness of our lives and worse detrimental to the goodness of our lives if it utilises too much of our cognitive and social resources for the acquisition of information that ultimately is of little or no axiological value (epistemic, ethical, or eudaimonic) for us personally and for society at large. Worse still, it can result, as we shall see in the chapters that follow (Chaps. 6, 7, 8, and 9), in the targeting and undermining of our autonomy, privacy, integrity and our digital identity by the Big Tech companies such as Facebook and Google, through their opaque and unaccountable algorithms that systemically target our information as users and data subjects, for their own self-seeking financial gain (Spence 2020).

Conclusion

The object of this chapter has been to explore how information generally and digital information specifically can be normatively evaluated in relation to the concept of a good life and well-being. To that end, the chapter developed and applied a meta-theoretical model, the DOIT-Wisdom model, as a groundwork for demonstrating how information can be normatively evaluated, using a triadic cluster of inter-related normative categories, those of the epistemic, ethical and eudaimonic categories, including the virtues that are integrally related to those axiological categories, analysed and defended on the basis of that model and on the basis of Stoic and neo-Stoic philosophy in Chap. 3. Central to those normative categories is the concept of wisdom, which being at once a meta-epistemological, meta-ethical and meta-eudaimonic concept, establishes a direct conceptual link between information and the notion of a good life and well-being both for individuals and societies generally. More specifically, wisdom as this chapter has demonstrated has a particular and important normative relevant role to play in the normative evaluation of digital information. The cited examples of the dissemination of digital information on the Internet, including other digital media and networks in Section "Wisdom and the Evaluation of Digital Information in Relation to a Good Life", were used illustratively as proof of concept to demonstrate how the notion of wisdom as the essential conceptual link between information, a good life and well-being can be used methodologically to evaluate axiologically (epistemically, ethically an eudaimonically) the creation, access, use and dissemination of any type of digital information on the Internet. Some digital informational evaluations such as, for example, the use of Skype or Zoom to communicate regularly with close friends, family and colleagues, will be positive, given that relationships with friends, family and colleagues are conducive to a good life and well-being and as such, their cultivation and maintenance, especially during COVID-19 pandemic, is generally, wise-worthy. Other digital informational evaluations, such as the examples from Facebook, SMS and Twitter above, may prove negative because not conducive to a good life and well-being and possibly conducive to the diminution of well-being, and therefore not wise-worthy. Although descriptively it can be said that we are now living in the age of information, we should act in a protreptic manner if we are wise or at least aspire to be wise, so as to promote the age of Wisdom, both for our own eudaimonic good and that of others, and especially for the sake of future generations who might mistake mere information and big data for knowledge and mere knowledge for wisdom. This chapter has argued for a theoretical groundwork for the normative evaluation of information generally, and digital information specifically through the notion of wisdom, a key principle of Stoic philosophy and generally Greco-Roman and Hellenistic philosophy. No doubt a lot more work must be accomplished in a more extensive examination and evaluation of specific types of good and bad uses of information that daily impact on the lives of people both positively and negatively. That is a topic we will examine in Chaps. 7 and 8 of this book. As we adapt ourselves to live more and more in the infosphere, including virtual worlds, this becomes a desirable and indeed a necessary task. I hope this is but the beginning of an ongoing research project.[23]

[23] See for example, Christopher Burr, Mariarosaria Taddeo and Luciano Floridi (2020). "The Ethics of Digital Well-Being: A Thematic Review," Science and Engineering Ethics, Vallor (2016). Technology and the Virtues.

References

Annas, Julia. (2011). *Intelligent Virtue*. Oxford: Oxford University Press.

Aurelius, Marcus. (1964). *The Meditations*, translated by Maxwell Staniforth. Harmondsworth: Penguin Books.

Aristotle. (1941). "Nichomachean Ethics", in Richard McKeon (ed.), *The Basic Works of Aristotle*. New York: Random House, 935–1112.

Baltes, PB. and Smith, J. (2004). "Lifespan Psychology: From Developmental Contextualism to Developmental Biocultural Co-constructivism." *Research in Human Development*, 1(3):123–144.

Baltes, PB., Gluck, J. and Kunzmann, U. (2002). "Wisdom: Its Structure and Function in Regulating Successful Life Span Development." in CR Snyder and SJ Lopez (Eds) *Handbook of Positive Psychology*, Oxford University Press.

Banicki, K. (2009). "The Berlin Wisdom Paradigm: A Conceptual Analysis of a Psychological Approach to Wisdom," *History and Philosophy of Psychology*, 11(2): 25–36.

Beyleveld, Deryck. (1991). *The Dialectical Necessity of Morality: An Analysis and Defense of Alan Gewirth's Argument to the Principle of Generic Consistency*. Chicago: University of Chicago Press.

Burr, Christopher, Taddeo, Mariarosaria and Floridi, Luciano. (2020). "The Ethics of Digital Well-Being: A Thematic Review," *Science and Engineering Ethics*, 26:2313–2343.

Descartes, R. (1979). "Principles of Philosophy," in *Philosophical Works of Descartes*, E. Haldane and G. Ross (trans. and eds.), London: Cambridge University Press, pp. 201–302.

Dretske, Fred. (1999). *Knowledge and the Flow of Information*. Stanford: CSLI Publications.

Elliott, Deni, and Spence, H. Edward. (2018). *Ethics for a Digital Era*. Oxford: Wiley Blackwell, Chapter 10.

Finnis, J. (1983). *Fundamentals of Ethics*. Oxford: Clarendon Press Oxford.

———. (1980). *Natural Law and Natural Rights*. Oxford: Clarendon Press Oxford.

Floridi, Luciano. (2005). "Is Semantic Information Meaningful Data?" *Philosophy and Phenomenological Research*, LXX:2, 351–370.

———. (2002). "What is the Philosophy of Information?" *Metaphilosophy*, 33, 123–145.

———. (1999). "Information Ethics: On the Theoretical Foundations of Computer Ethics." *Ethics and Information Technology*, 1(1), 37–56.

Garrett, R. (1996). "Three Definitions of Wisdom." In Lehrer and Smith. (eds.), *Knowledge, Teaching, and Wisdom*. Dordrecht: Kluwer Academic Publishers, pp. 221–232.

Gewirth, Alan. (1996). *The Community of Rights.* Chicago, University of Chicago Press.

———. (1998). *Self-fulfillment.* Princeton, NJ: Princeton University Press.

———. (1978). *Reason and Morality.* Chicago: University of Chicago Press.

Kekes, John. (1995). *Moral Wisdom and Good Lives.* Ithaca: Cornell University Press.

Kekes, John. (1983). "Wisdom", *American Philosophical Quarterly*, 20(3): 277–286.

Kvanvig, J. L. (2003). *The Value of Knowledge and the Pursuit of Understanding.* Cambridge: Cambridge University Press.

Kunzmann, Ute and Baltes, Paul. (2005). "The Psychology of Wisdom: Theoretical and Empirical Challenges", in Robert J. Stenberg and Jennifer Jordan (eds.), *A Handbook of Wisdom: Psychological Perspectives.* New York: Cambridge University Press.

Lehrer, Tom and Smith, Chris. (1996). *Knowledge, Teaching and Wisdom.* Dordrecht: Kluwer Academic Publishers.

Maxwell, Nicholas. (1984). "From Knowledge to Wisdom: Guiding Choices in Scientific Research". *Bulletin of Science, Technology and Society.* 4 (4):316-334.

Maxwell, N. 2007. *From Knowledge to Wisdom* (Second Edition), London: Pentire Press.

Nozick, R. (1989). "What is Wisdom and Why Do Philosophers Love It So?" *The Examined Life.* New York: Touchstone Press, pp. 267–278.

Plato. (1978). The Apology, in the Collected Dialogues of Plato. Edith Hamilton and Huntington Cairns (eds.) Princeton: Princeton University Press, pp. 575–844.

Ryan, Sharon. (1996). "The Epistemic Virtues of Consistency." *Synthese*, 109 (2): 121–141.

Ryan, S. (2014). ""Wisdom". *Stanford Encyclopedia of Philosophy.* Accessed May 30, 2025.

Ryan, S. (1999). "What is Wisdom"? *Philosophical Studies*, 93: 119–139

Spence, Edward H. (2020). "The Sixth Estate; Tech Media Corruption in the Age of Information." *Journal of Information, Communication and Ethics in Society*, 18(4):553–573. Bingley, UK: Emerald Publishing Limited, DOI: https://doi.org/10.1108/JICES-02-2020-0014.

Spence, Edward H. (2009). "A Universal Model for the Normative Evaluation of Internet Information", *Ethics and Information Technology*, 11(4): 243–253.

Spence, Edward H. (2011). "Information, Knowledge and Wisdom: Groundwork for the Normative Evaluation of Digital Information and Its Relation to the Good Life," *Ethics and Information Technology*, 13(3): 261–275.

Spence, Edward H. (2006). *Ethics Within Reason: A Neo-Gewirthian Approach.* Lanham: Lexington Books (a division of Rowman and Littlefield).

Tiberius, V. (2008). *The Reflective Life: Living Wisely With Our Limits.* Oxford: Oxford University Press.

Vallor, S. (2016). *Technology and the Virtues: A Philosophical Guide to a Future Worth Wanting*, Oxford: Oxford University Press

Varelius, J. (2004). "Objective Explanations of Individual Well-Being." *Journal of Happiness Studies* 5: 73–91.

Vitec, B. and Jackson, W. (2008). *The Virtues of Ignorance: Complexity and Sustainability, and the Limits of Knowledge.* Lexington, Kentucky: The University Press of Kentucky.

CHAPTER 3

Wisdom and Control: Applying Stoic Philosophy to Technology

O my friend, why do you who are a citizen of the great and mighty and wise city of Athens, care so much about laying up the greatest amount of money and honour and reputation, and so little about wisdom and truth … Are you not ashamed of this?
—Plato (Apology, 1978)

Introduction

In the last few years, there has been a notable and steady revival of Stoic philosophy. Scholarly works by A. A. Long, Gisela Striker, Julia Annas, Pierre Hadot, Martha Nussbaum, Brad Inwood, John Sellars and Laurence Becker,[1] to name but a few, have generated a renewed interest in Stoic

[1] A. A. Long, Hellenistic Philosophy: Stoics, Epicureans, Sceptics (1986, 2nd-edn), Striker, Essays on Hellenistic Epistemology and Ethics (1996), J. Annas, The Morality of Happiness (1993), M. C. Nussbaum, The Therapy of Desire (1994), and L.C. Becker, A New Stoicism (1998); John Sellars, Lessons in Stoicism: What Ancient Philosophers Teach Us about How to Live (2020); John Sellars, Stoicism (2020); Brad Inwood, Stoicism: A Very Short Introduction (2018); Pierre Hadot, The Present Is Our Happiness (2011); Pierre Hadot, The Inner Citadel: The Meditations of Marcus Aurelius (2001); Pierre Hadot, Philosophy as a Way of Life (1995), Robertson, Donald. How to Think Like and Emperor: The Stoic Philosophy of Marcus Aurelius. (2019)

E. H. Spence, *Wisdom in the Age of Intelligent Machines*,
https://doi.org/10.1007/978-3-031-93883-2_3

philosophy and especially in Stoic ethics. Outside academic philosophy, Stoicism has also been exerting an indirect influence. I am referring to the remarkable rise of public philosophy, both offline and online, and practical and professional ethics, over the last 20 years or so. Professional ethics has been growing exponentially in almost all areas of professional practice, as in nursing, medicine, policing, engineering, computers, technology and the media. With regard to public philosophy, the dissemination and discussion of philosophical ideas in the context of issues relating to day-to-day living has also been taking place in public venues such as cafes, restaurants, theatres, pubs, vineyards, art galleries and even opera houses, across the world. Organised by professional and academic philosophers, these public philosophy forums have proved enormously popular with members of the general public, most of who have no prior philosophical background.[2] Generally, Hellenistic in spirit because of their practical orientation as a way of life, public philosophy and, to a lesser extent, practical and professional ethics have a Stoic pedigree, because of the robust professional, political, social and communal engagement that characterises them both.[3]

[2] In 1997, Edward H. Spence founded the Philosophy Plays Project. Since then to the present time (2021) he has produced and directed, in the spirit of Platonic Symposia, a program of public philosophy in Australia through the medium of philosophy plays, that combine philosophical talks by academic philosophers with dramatic performances by actors. Several of the philosophy plays written by Spence have been performed in various Arts Festivals in Australia. His philosophy play, The Philosophy of Love: Love in the Age of Terror, was performed at the Sydney Opera House in 2004; his Stoic philosophy play, Zeno's Secret: How to be Happy in an Unhappy World, was performed at the Factory Theatre in 2018 for the Greek Festival of Sydney and in 2021 his philosophy play The Last Freedom: Algorithms and Ataraxia, a play based on the theme of this book on Stoic Philosophy and Technology, will be performed for the Greek Festival of Sydney at the Art Hellenic Theatre. The ultimate aim of the project is to cultivate and socialise philosophy by raising public awareness of its practical application and usefulness as a "way of life" in helping us, as individuals and collectively as communities and society generally, to think more critically and reflectively about how we live our lives. In Martha Nussbaum's apt phrase, taken from the title of her book, it is a way of "cultivating humanity" (Nussbaum 1997).

[3] For Stoicism in public philosophy and professional ethics see, E. H. Spence (2004), "Philosophy Plays: A Neo-Stoic Method for Teaching Ethics," (2004), and E. H. Spence "The Theatre of Philosophy: A Neo-Socratic Method for Performing Public Philosophy," (2012).

Unlike Epicureanism, Stoicism encourages active political and social engagement and not merely a retreat to a communal garden; unlike Cynicism, Stoicism considers preferred indifferents like health, wealth and social status as desirable so long as they accord with a virtuous lifestyle. The Stoics, however, concur with the Cynics that only virtue is good and that it alone is both necessary and sufficient for a good and happy life, a eudaimonic life. Unlike the Sceptics, the Stoics do not suspend judgement on all issues concerning the natural, political and social environment they find themselves in. On the contrary, they counsel the active and rigorous pursuit of rational knowledge both of oneself and one's natural, social and political environments as a necessary condition for acquiring practical wisdom through virtue for the attainment of eudaimonia, the ultimate goal for the Stoics. Thus, the practical orientation, the vigorous engagement in the political, social and communal life of one's cultural and natural environments, both locally and globally, as well as the rigorous pursuit of rational knowledge as a necessary prerequisite in attaining practical wisdom essential for virtue and self-fulfilment, are central characteristics of Stoic philosophy and in particular Stoic ethics. In that regard, the proliferation of professional ethics and public philosophy in the pursuit of professional ethical knowledge and practical wisdom, even if not directly influenced by Stoicism is, at the very least, in terms of orientation and approach, attuned to some of its central precepts and principles. And that suffices to show that the practice of professional ethics and public philosophy is very much in keeping with the spirit and rationale of Stoicism, and more generally, Ancient Greek and Roman philosophy.

Stoicism also shares with other Hellenistic schools of philosophy, Epicureans, Cynics and Sceptics, the central notion that philosophy has to be practical; for it is only if it is practical that it can prove valuable in helping people to live good lives both as individuals and as societies for the attainment of eudaimonia or happiness. This notion is similar to one raised by Plato in The Republic, a notion that in all probability influenced the Stoics, namely that justice requires both personal and collective societal ethical harmony. It is perhaps that idea that informed and encouraged Stoics like Seneca and the Emperor Marcus Aurelius, to devote themselves to public life, although in the case of Seneca, as with Socrates before him, it cost him his life. There is always danger, fatal in Seneca's case, when a sage aligns himself to a serial killer. As Plato himself discovered with regard to the King of Syracuse, tyrants are not always amenable to the self-abstaining rigors of an ethical lifestyle.

The above preamble was intended primarily to indicate the tangible continuing legacy of Stoic philosophy and in particular ethics in contemporary practical and professional ethics, as well as in both scholarly and public philosophy. In what follows I will demonstrate that Alan Gewirth's rationalist ethical theory, in all its different facets, especially as set out in his book *Self-fulfillment* (1998), is paradigmatically and essentially Stoic in its structure, scope, rationale and vision, and schematically in its content. Moreover, I will show that Gewirth's ethical theory helps clarify and rationally support some of the central doctrines of Stoic ethics in a way that renders those doctrines more compelling and persuasive for a contemporary audience.

To demonstrate the essential Stoic features of Gewirth's ethical theory I will refer extensively to relevant passages in Gewirth's book, Self-fulfillment (1998), which bears close similarities to Stoic philosophy and ethics in particular, especially as regards the Stoic views on eudaimonia or well-being, virtue and reason. I will argue that according to my original interpretation and analysis, Gewirth's notion of self-fulfilment is a neo-Stoic version of Stoic eudaimonia.[4]

One key doctrine of Stoic ethics, which may first appear to be at odds with Gewirth's own views on the matter, is the controversial Stoic claim that virtue alone as the only good is both necessary and sufficient for happiness or eudaimonia (in this chapter, I will use those terms as well as that of well-being, interchangeably). At the outset, Gewirth appears to be more Aristotelian on this issue, who believed that though virtue was necessary for eudaimonia, it was not sufficient. However, there is a neo-Stoic way of understanding the necessity as well as the sufficiency of virtue for happiness, one that Laurence Becker convincingly expounds in his book A New Stoicism (1998), which, I believe, accords well with my analysis of the neo-Stoic view, which I am ascribing to Gewirth in this chapter.

If I am right, the neo-Stoic character of Gewirth's ethical theory, to whom I am ascribing on the basis of my neo-Stoic reconstruction of his theory of self-fulfilment, is at once both surprising and yet not: surprising because Gewirth himself has not consciously or overtly recognised, as far as I am aware, the pronounced and crucial Stoic character of his ethical

[4] See Spence, E. Ethics Within Reason: A Neo–Gewithian Approach. (2006), pp. 393–442.

and eudaimonic theory.[5] Although frequently making references to the Aristotelian and Kantian elements in his own theory,[6] he rarely if ever makes overt or even indirect references to similarities or differences between his theory and that of the Stoics.[7]

Before proceeding to examine the close relationship between the Gewirthian neo-Stoic theory I am ascribing to Gewirth, and the traditional ancient philosophy of Stoicism, I will first provide a brief historical summary of some of the central characteristics of Stoic philosophy. A further closer and systematic analysis of those Stoic principles would be provided through my comparative analysis of Stoic philosophy and the parallel Gewirthian neo-Stoic theory that follows.

Stoic Philosophy

Most of what is known of Zeno's philosophy, the founder of Stoic philosophy, is through the writings of later Stoics, particularly the Roman Stoics, Epictetus, Seneca and the Emperor Marcus Aurelius. Zeno, like all the Stoics, preached equality of the sexes, and also claimed that men and women should dress alike. Observing how unisex fashion has gained ascendancy, one could say that Zeno and his Stoics were in truth the early trendsetters. According to the Stoics, philosophy is like an animal, logic corresponding to the bones and sinews, ethics to the fleshy parts, physics to the soul. Though Stoic writings, especially of the early Stoics, including Zeno, Cleanthes and Chrysippus, covered all topics including logic, theory of knowledge (Epistemology), metaphysics and ethics, it is the latter that has had the biggest and more profound surviving influence, mainly

[5] When in conversation with Gewirth (Chicago, March 2001), I raised the issue of the striking similarity between his ethics and that of the Stoics, he expressed both interest and surprise. I hope to show that the surprise for Gewirth should be a pleasant one and the interest is both warranted and justified.

[6] For a comparison of Kant's and Gewirth's ethical theories, see Spence E., A Commentary on Deryck Beyleveld's "Gewirth and Kant on Justifying the Supreme Principle of Morality." In Michael Boylan (ed.), Gewirth, Critical Essays On Action, Rationality, And Community (1999).

[7] See especially his *Self-fulfillment* (1998). Although references to Aristotle abound, and the ones to Kant are not infrequent, the references to the Stoics are almost conspicuously scarce, if not entirely absent.

through the works of the later Hellenic and Roman Stoics, notably those of Epictetus, Seneca and Marcus Aurelius. Given also, its contemporary importance and relevance, it is on ethics that the following exposition of Stoic philosophy will primarily focus.

The Life of Virtue and the Life of Wisdom

According to the Stoics, the life of virtue is both necessary and sufficient for living a happy life. Essential to understanding this, is the distinction the Stoics made between those things that are completely within our power and control and those things that are not always and not completely within our control. According to Epictetus (2018),

> Our opinions are up to us, and our impulses, desires, aversions – in short, whatever is our doing. Our bodies are not up to us, nor our possessions, our reputations, or our public offices, or, that is, whatever in not our doing… So remember, if you think that things naturally enslaved are free or that things not your own are your own, you will be thwarted, miserable, and upset, and will blame both gods and men. (Enchiridion, Chapter 1:1–3)

The only thing over which we have control, according to the Stoics, is our faculty of judgement. Since anything else, including all external affairs and circumstances are not within our power, we should adopt towards them an attitude of indifference. According to Epictetus (2018),

> What upsets people is not things themselves but their judgments about things. For example, death is nothing dreadful (or else it would have appeared so to Socrates), but instead the judgement about death is that it is dreadful, that is what is dreadful. (Enchiridion, Chapter 5)

To avoid unhappiness, frustration and disappointment, we therefore need to do two things: control those things that are within our control (our beliefs, judgements, desires and attitudes) and be indifferent to those things which are not in our control (things external to us). As rational beings, we should therefore perfect our characters through living a virtuous life because it is the only thing that can bring us both ataraxia (tranquillity), autarkia, (self-sufficiency) and eudaimonia (happiness); a happiness that is totally within our control, just because it depends on our own judgement, which is entirely within our control.

Associated with the idea of a virtuous life is the idea of wisdom. The ultimate object of philosophy according to the Stoics is to teach us not knowledge, but wisdom, understood as a way of living a virtuous life. Wisdom, therefore, understood in the Stoic sense was a way of life that brought peace of mind (ataraxia), inner freedom (autarkia), happiness, (eudaimonia) and a cosmic consciousness. By "cosmic consciousness," the Stoics understood that quality, as universal reason that permeates the whole cosmos and by virtue of which we all are, as human beings, integrated parts of the cosmos. As Pierre Hadot tells us (1995),

> The exercise of wisdom entails a cosmic dimension. Whereas the average per son has lost touch with the world, and does not see the world qua world, but rather treats the world as a means of satisfying his desires, the [Stoic] sage never ceases to have the whole constantly present to mind. He thinks and acts within a cosmic perspective. He has a feeling of belonging to a whole, which goes beyond the limits of his individuality.

Most importantly, the Stoics viewed philosophy as therapeutic, intended to cure mankind's anguish. For according to an Epicurean saying, "vain is the word of the philosopher which does not heal any suffering of man" credited to an inscription erected by Diogenes of Oinoanda, province of Lycia, in 120 AD to honour Epicurus.

Social and Political Engagement

Closely associated with the cosmic perspective in Stoic philosophy, is the communal perspective. That is, the concern for living in the service of human community, and for acting in accordance with social justice. Thus, according to the Stoics, philosophy entails a community engagement. This is something that the Stoics have in common with Aristotle who saw philosophy as essentially political. That is, concerned with the affairs of the polis. For the Stoics, however, the polis was the cosmopolis, not the city-state but the whole world.

Significantly, what is missing in most people's lives according to the Stoics is the practice of wisdom understood not as the acquisition of mere knowledge or information but as a way of being in the world that requires one to live a life in accordance with virtue. More than information, wisdom requires transformation. More than knowledge, wisdom requires practical ethics. Some knowledge about the world, even abstract and theoretical knowledge is necessary for wisdom, but it is not sufficient. To become wise, one requires not only to be informed but more importantly to become transformed through the practice of philosophy, specifically Stoic philosophy, as a way of life. This is by no means an easy task and it may in fact prove very difficult for most people. However, the threat to the natural environment through continuous commercial exploitation to satisfy our insatiable desire for consumer goods as well as the constant outbreak of wars around the globe, not to mention, the current and ongoing pandemic across the world, in 2020–2021, renders the ethical lifestyle recommended by the Stoics, with its emphasis on a simple, peaceful and a frugal but happy way of living, that respects not only the ethical rights of all individuals as citizens of the world but also extends that respect to the whole of the natural environment, as well as the digital environment, as we shall examine later in Chap. 4, as ethically desirable.

Philosophy as a Way of Life

The Stoics viewed philosophy not just as an interesting intellectual exercise, but as a way of life available to everyone—the Art of Life or "techne viou." The following is an outline of the reasons of why the Stoics viewed philosophy in this way:

(1) The goal in life is to be happy, for who would not want to be happy.
(2) Only the possession of what is good can make us truly happy.
(3) Moreover, only the good that is always within our control can make us happy.
(4) The only good is virtue and moreover virtue is always within our control. For being virtuous, is always up to us. The virtues, among others, sometimes referred to as the cardinal virtues, are justice, temperance or moderation, prudence, or practical wisdom and courage.
(5) Therefore, only the possession of virtue can truly make us happy.
(6) The possession of other things like health, wealth, fame, beauty, status, popularity are desirable but are not always good. Their possession can sometimes bring us unhappiness, not happiness. Moreover, their possession can sometimes result in unethical and bad conduct that harms others and causes their unhappiness.

(7) Hence, although in normal circumstances things like wealth and status should be preferred, their possession is unnecessary for happiness. They are merely preferred indifferents.
(8) Therefore, a virtuous person can be happy without them since only virtue, as the only good, is what one requires to be happy.
(9) But why is virtue the only good?
(10) Virtue is the only good because only virtue is unconditionally good. For in all circumstances, virtuous conduct is in agreement with nature, and it alone determines what is good.
(11) Hence, virtue is the only good because it allows one to live in accordance with nature.
(12) But what does to live in agreement with nature, mean?
(13) Living in agreement with nature means simply, living a life in accordance with reason for nature is perfectly rational.
(14) Therefore, a virtuous life is a life lived in accordance with the perfection of one's reason.
(15) And moreover, reason is something that we share with all of nature and God, since God and nature are, according to the Stoics, one and the same, since the Stoics identified God with universal reason that permeates the whole of nature.
(16) Thus, in living rationally and virtuously we share in God's perfection.
(17) And insofar as philosophy helps us to live a life in which we share in God's perfection, then philosophy is truly the art of life.

Philosophy can be viewed as the art of living in two ways: (a) as a craft or art, which once learned can enable one to live a good and fulfilling life on a daily basis and (b) the perfection of one's human nature—becoming the best of one's kind—humankind. For the Stoics, the perfection of one's human nature meant the perfection of one's rational nature. For according to the Stoics, the essence of human nature and nature generally, the whole universe in fact, was reason. Thus, in perfecting one's rational nature one becomes at once the object and the subject of art, that is, the art of life. This was the highest form of art, for in perfecting one's rational nature through the practice of philosophy one became like God since God, as we saw, was identified by the Stoics with nature, which they viewed as supremely rational and orderly. In conclusion of this section, philosophy can, according to the Stoics, be conceived as the art of living in these two ways: as a process, an acquired and learned craft whose application can enable one to live happily regardless of the slings and arrows of outrageous fortune; and as a product or an end, namely, the perfection of one's rational nature which we share with God. In this second idealistic and aspirational sense, one can become through the perfection of one's rational character, godlike.

As a point of interest, the story of creation in the Book of Genesis (King James Bible, Chapter 3:22), viewed here as an allegory, tells of God expelling Adam and Eve from Eden, as a precautionary measure "lest they eat from the Tree of Life and become like God immortal." The suggestion here in the form of an allegory is perhaps humanity's potential in becoming through wisdom in the form of existential meta-knowledge of how to live one's life, immortal and godlike. We shall return to that suggestion in the final chapter of this book, Chap. 8, by exploring the possibility of a beneficial symbiotic relationship, between humans and artificial general intelligent, AI agents. The Stoic idea is that humans, through virtue and living in accordance with nature perceived by them to be divine, can achieve perfect eudaimonia and become part of God. Though conceived differently, the Christian idea of salvation and redemption of humankind after the fall, through Jesus Christ's divine sacrifice, and the Stoic idea of becoming part of God or nature by the perfection of oneself through virtue, though different in content and context, appear to run parallel, around the central idea of humanity's potential for becoming godlike and sharing God's kingdom: in the Bible story, God's Heaven, and in the Stoic story, this Natural World, in which God, whom the Stoics identified with nature, is immanent, ever-present, omnipotent and benevolent. The characteristics traditionally applied by theist philosophers to the Christian God.

The Stoic Relationship of Happiness to Virtue

Another central feature of the Stoic conception of happiness is its intrinsic relation to virtue. For as noted earlier, virtue as the primary comprehensive goal of human nature is not only the essential means for the attainment of happiness, but more importantly, it is constitutive of it (Long 1986, 197). In being the end or goal of life, virtue is both the means and the end of a happy life. The means, for the development and inculcation of virtue leads to happiness; and the end, because its possession as the only good guarantees a good life; for virtue is constitutive of happiness and its possession as the only good necessarily guarantees one's happiness. Another related characteristic of happiness is its self-sufficiency; for insofar as virtue as the only good is entirely within one's control, and that alone is sufficient for happiness, then happiness results from virtuous self-sufficiency (Long 1986).

An important qualification, however, must be made to the view that happiness and virtue in Stoic ethics are so intimately and intrinsically connected, that one cannot tell them apart, and as a result, one might then be led to view Stoic happiness as nothing more than a life of virtue. This would be a mistake, however. For as Becker correctly points out, Stoic virtue is not the same as say Kantian duty (Becker 1998). There is joy in Stoic ethics over and above the state of just being virtuous and always doing the right thing out of a grudging sense of duty. In Stoicism, the possession of virtue f ills one's soul with joy similarly to how a state of health and fitness fills one with a general sense of well-being, both physically and psychologically. The Stoic sage, as the sage in Plato's Symposium standing at the top of the ladder of human perfection and gazing at the form of the good, is in love with the good[8] and that alone fills him with abundant joy.

Julia Annas (1993), rightly insists, contrary to the view that Stoic eudaimonia is nothing more than a life of virtue, that we must take Stoic eudaimonists at their word for ultimately what they are offering us is an account of a happy life and not just a virtuous one. And these two things are, though intrinsically connected, conceptually distinct, as are shape and size, for example, although shape and size are intrinsically connected, they are nevertheless, conceptually distinct. Insofar as virtue is the only good and virtue is entirely within one's control, then possession of the only good is both necessary and sufficient for happiness and potentially, within the reach of everyone. Moreover, sages like Socrates and the Buddha are real exemplars that render this view not a mere philosophical abstraction but a real practical possibility.

[8] In her review of Martha Nussbaum's Cultivating Humanity: A Classical Defense of Reform in Liberal Education (1997), Marilyn Friedman points out that the "idea of love for all humanity," is mentioned by Nussbaum "no less than fourteen times" (Cultivating Humanity, pages 6, 7, 13, 14, 36, 61, 64, 67, 72, 84, 103, 222, 259, 292)—see Marilyn Friedman, "Educating for World Citizenship" Ethics (April 2000): 589. Nussbaum's "love for all humanity" seems essentially similar to the "love of the Good," which, as the primary motivation for doing what is right, I attribute to both the Stoic and the neo-Stoic.

Stoic Virtue as Both Necessary and Sufficient for Happiness

From the two propositions that human virtue is rational perfection, and that the perfection of one's reason is in agreement with nature, the Stoics inferred that virtue is good, the only good, for it alone allows one to live in agreement with nature (Striker 1992). Moreover, being a function of human rationality virtue is within one's rational control and as the only good, its possession leads to happiness or eudaimonia. That virtue is the only good follows from the view that only virtue is unconditionally good. For in all circumstances, virtuous conduct in agreement with nature, alone determines what is good. Hence, virtue is the only good because it allows one to live in accordance with nature.

Whereas health, wealth, friends and social status might be good, at least some, or most of the time, what the Stoics referred to as "preferred indifferents" (the Stoic term for contingent "goods"), they are not always good, as their possession may not result in a virtuous life in agreement with nature. Worse still, they may lead one to unethical and vicious actions that are bad because such actions are not in agreement with nature; on the contrary, they are in disagreement with nature because, ultimately, they are inconsistent with one's rational nature.

The above analysis is, I believe, in keeping with the general features of Stoic ethics concerning the sufficiency of virtue for happiness.[9] What emerges is the following schema:

(1) Virtue is the perfection of one's rational nature.
(2) Virtue as the perfection of one's rational nature allows one to live in agreement with nature.
(3) As the only thing that allows one to live a life in agreement with nature, virtue is the only good.
(4) As such, virtue is unconditionally good.
(5) Only possession of what is unconditionally is good can render one happy.
(6) Therefore, since only virtue is unconditionally good, only virtue can make one happy.
(7) Virtue is completely within every rational person's control.

[9] In my analysis, I have followed closely Laurence C Becker's insightful neo- Stoic and Stoic analysis on the sufficiency thesis of Stoic eudaimonia (ANS:108–158)

(8) Hence, every rational person can potentially (because everyone is by nature inherently rational) become happy and, moreover, must aim at becoming happy by becoming virtuous.

The above eight-point outline renders, I hope clearly, the intrinsic relationship between the central features of Stoic ethics: reason, virtue, nature and eudaimonia. Moreover, the outline draws attention to both the descriptive as well as the prescriptive characteristics of Stoicism. Furthermore, it highlights the ideal and developmental aspects of Stoic ethics. However ideal, the status of sagehood, as the only possible state for true and lasting happiness, remains a practical possibility for every rational person. In the Tusculan Disputations, Cicero devotes the whole book (V) to a defence of the necessity and sufficiency of virtue for happiness. Laurence C. Becker has identified the following premises as the key premises in Cicero's argument (Becker 1998):

(1) "No one can be happy except when good is secure and certain and lasting (V.xiv.40).
(2) The good of virtue is secure and certain and lasting because (a) once achieved, its maintenance is within the agent's control (V.xiv.42), and (b) it is free from the disturbances of the soul that produce wretchedness (V.xv.43).
(3) Moreover, in its affective dimension, a virtuous life is characterised by tranquillity and joy and thus may unproblematically be described as a happy life (V.xv.43)
(4) No form of happiness can be good unless it includes, or is founded upon, virtue, or what is right" (V.xv.44–5).

Gewirth: Definition of Virtue

It will be recalled that a Stoic definition of virtue is that "it is not defined by the consequences in the world which it succeeds in promoting but by a pattern of behavior that follows necessarily form a disposition perfectly in tune with Nature's rationality" (Long 1986). This definition of Stoic virtue is very much in keeping with Gewirth's own understanding of virtue: namely, a "disposition perfectly in tune with human rationality" where human nature as rational purposive agency replaces the metaphysical notion of Stoic Nature. And for Gewirth, so too for the Stoics, "the right thing to do is that which accords with virtue, and this is equivalent to saying that it accords with the nature of a perfect rational being (emphasis added)" (Long 1986). For Gewirth, the right thing to do accords with the requirements of the PGC, which, in turn, accords with the natural property of "rational purposive agency." That, in turn, requires the development and inculcation of the virtues as the enabling dispositions for acting and living ethically. And acting and living virtuously then accords, at least for the neo-Stoic sage, with the nature of a rational purposive agent. And being a common natural property that all rational agents have in common, rational purposive agency is therefore universal.

Gewirth's Relationship of Happiness to Virtue

Gewirth's attention to the virtues and their relationship to both rationality and morality, mediated through the PGC, is extensive and comprehensive. What will concern us in this section is Gewirth's account of the relationship between virtue and self-fulfilment. The relationship as expressed by Gewirth appears to be as follows: insofar as morality is intrinsically related to self fulfilment as capacity fulfilment (Gewirth 1998), and insofar as the virtues are essential in enabling rational agents to act and live ethically in compliance with the requirements of the PGC (Gewirth 1998), then the virtues are also intrinsically related to self-fulfilment (Gewirth 1998). The above-described relationship between virtue and self-fulfilment appears, in the first instance, to be instrumental. That is, virtues are essential because they are effective means for ethical development thus enabling one to comply more effectively with both the negative and positive requirements of the PGC.

This instrumental understanding of the virtues is also part of the developmental aspect of Stoic ethics. By aiming and attempting to act and live virtuously, one gradually becomes inculcated in virtuous conduct through habituation or *oikeiosis*. The Stoic concept of oikeiosis will be discussed in more detail below.[10]

However, there is also, as in Stoic ethics, a constitutive aspect of the virtues in Gewirth's model of self-fulfilment. Once one becomes habituated to acting ethically in compliance with the PGC, then virtuous

[10] For a detailed discussion of Stoic Oikeiosis and its structural similarity with a similar concept used by Gewirth in his ethical theory, see E. Spence, Ethics Within Reason: A Neo-Gewithian Approach (2006), pp. 433–442.

conduct becomes, as it were, one's second nature. In the Gewirthian neo-Stoic sage, at the limit of virtuous perfection, however, virtue becomes, together with rationality, the sage's first and only nature. The perfection of reason and virtue combine and match in the neo-Stoic sage, and though remaining conceptually distinct, become in the sage's actions, practically indistinguishable. Thus, the neo-Stoic sage acts virtuously not only because his reason demands it, but more importantly because his reason is attuned to virtuous conduct. Unlike the Kantian "sage," the Gewirthian neo-Stoic sage, like his Stoic counterpart, does the right thing out of love for the good and not out of a joyless sense of duty. The neo-Stoic sage, like his Stoic predecessor, enjoys being virtuous and this, unlike the Kantian sage, does not in any way diminish his goodness or his happiness.

The *developmental* and the *constitutive* aspects of virtue are illustrated by Gewirth in the following passage: "There is a difference, however, between one's having a duty to act form a virtue and one's having a duty to try to develop a virtue. It is the latter that is required for capacity-fulfillment" (Gewirth 1998). When, however, one becomes self-fulfilled, by becoming perfectly virtuous qua a rational person, then one's developmental virtue becomes constitutive both of his character and of his happiness or self-fulfilment. Acting out of virtue for a Gewirthian neo-Stoic sage, as for the Stoic sage of antiquity, is acting out of his perfected rational nature. There is, in contrast to the Kantian "sage," no separation in the Stoic sage between the two distinct thoughts "this is what I want to do" and "this is what I ought to do." Acting out of desire and acting out of virtue become in the neo-Stoic and Stoic sage alike one and the same thing.

Gewirthian Virtue as Both Necessary and Sufficient for Happiness

Insofar as health, wealth, fame and social status, some of the Stoic-preferred indifferents are not essential for complying with the requirements of the PGC, and moreover, under certain circumstances their possession may result in the violation of those requirements, then virtue would appear to be, as for the Stoics so too for Gewirth, the only unconditional good. Unconditional, because only virtue is under all circumstances, at least in principle, capable of enabling a rational purposive agent to act in accordance with the negative and positive requirements of the PGC. I say only capable, because like rationality, virtue cannot motivate a sociopath or a psychopath to behave ethically. This is partly because the sociopath's and the psychopath's rational nature has, in some way, become dysfunctional. But even if virtue as the only good, understood as the only unconditional good that is capable of enabling a rational purposive agent to act and live ethically is necessary for self-fulfilment, is it sufficient?

In my book, *Ethics Within Reason: A Neo-Gewithian Approach*, Lexington, (Spence 2006), I explore in detail that question and conclude that as in Stoic philosophy so too in the neo-Stoic philosophy I am attributing to Gewirth in this chapter, virtue is indeed both necessary and sufficient for both neo-Stoic eudaimonia and Stoic eudaimonia. Constrained by space, I am not able to include in this chapter the full detailed analysis I provide in my book (Spence 2006). However, I will provide in summary an outline of the conclusion of the argument I reach in my book. Namely, that constitutive virtue[11] is sufficient for ideal self-fulfilment and equivalently eudaimonia in Gewirthian neo-Stoic philosophy so long as a minimal rational purposive agency can be preserved and maintained.[12] This will, in turn, require the presence of the material, personal and social conditions that are sufficient for the preservation and maintenance of rational agency, nothing more and nothing less. An interesting contrast that seems to support the social dimension of the neo-Stoic sagehood I have outlined above is a passage from Julia Annas: "virtue is sufficient for non-unhappiness, but for happiness the person must be able to use, not merely possess, virtue" (1999).

[11] To remind us of what was said earlier in regard to virtue, Gewirth makes a distinction between developmental and constitutive virtue. Developmental virtue is what enables one to become ethically a good person. Its function is primarily instrumental as physical fitness is for health. However, once one becomes ethically good, in the sense that one becomes disposed to always act out of virtue, which at this stage of one's development has become identical with one's rational purposive agency, virtue becomes constitutive of that person's ethical goodness. To use the same previous analogy, one's physical fitness is no longer merely instrumental for becoming healthy, but it becomes, once one attains and maintains good health, constitutive of one's health and general lifestyle. When constitutive, virtue, like physical fitness, is its own reward—it becomes together with moral goodness an end in itself and not merely an instrumental means for ethical conduct.

[12] The necessity of purposive agency as a prerequisite for Stoic, and in particular, neo-Stoic eudaimonia or happiness is also alluded to by M. Nussbaum, in Symposium on Cosmopolitanism, "Duties of Justice, Duties of Material Aid: Cicero's Problematic Legacy," (2000), 191.

Oikeiosis

The Stoics[13]

The transition from developmental to constitutive virtue, from imperfect and incomplete virtue to perfect and complete virtue, requires a kind of ethical transformation. In an insightful passage, Julia Annas describes quite clearly the kind of ethical transformation that Stoic happiness requires:

> As a theory of happiness [the Socratic theory of happiness in Plato], it is very demanding. You won't have got your account of happiness right until you achieve the right perspective, and that will be utterly different from your initial standpoint. Becoming virtuous requires *transforming* (emphasis added) your values and priorities; nothing short of this will do. All your valuations of conventional goods and evils will turn out to be wrong ... we are used to moral theories that are demanding, but it is not easy for us to conceive of such a theory which is a theory of happiness, rather than something else. But this is not an argument against a theory like Plato's. It shows merely that it takes a long intellectual journey away from common sense and conventional beliefs to achieve the wisdom of the virtuous person and thus to have your whole life, and your conception of happiness transformed. (Annas 1999)

Gewirth also recognises and acknowledges the transformative aspect of his concept of ideal self-fulfilment. As he tells us, "to recognize and accept these requirements in one's thought and action may involve a kind of self-transformation (emphasis added) whereby one becomes more fully a moral and social being" (Gewirth 1998).

It is as difficult for the Stoic and neo-Stoic sage to accept the common-sense view of happiness, primarily the hedonistic view of happiness, as it is for the ordinary person to accept the Stoic and neo-Stoic view of happiness as a view of "happiness." And the difficulty each has in comprehending the attractive ness of each other's notion of happiness lies in their vastly different ethical perspectives. It is for this reason that a practical and gradual transformation must take place over a long period of a person's life

[13] Stoics: Becker, A New Stoicism, 56; 74; 90; 93–95; Long, Hellenistic Philosophy, 172–74; 185–88; Striker, "Stoicism," 1209–1210, and Essays, 178;181; 219; 225–26; 250–51; 253–54; 257–261; 267; 281–297. 31. Self-fulfillment, 45;60;65;86; 95–96;112;123–35;222–23;226.

before a hedonistic perspective on happiness can be transformed and replaced in the person aspiring to Stoic or neo-Stoic sagehood, with a eudaimonic perspective of happiness.

It is the realisation of the practical necessity for such transformation and the need for providing a conceptual account of how such a transformation can practically take place, that the Stoics developed their celebrated theory of *oikeiosis*; a theory which is considered to be one of the most crucial and fundamental doctrines of Stoic ethics. The theory purports to show how a per son can progressively develop, because he is naturally conditioned and thus capable of doing so, from the state of having and being motivated by purely hedonistic desires as a child, to the state of acquiring, over a time period of biological, psychological, social, but more importantly rational development, an ethical perspective.

Gisela Striker defines "oikeiosis" as "appropriation" or "familiarisation" with oneself and others (Striker 1992), or as "recognition and appreciation of something belonging to one" (Striker 1996). In a similar vein, A.A. Long defines it as the disposition of being "well-disposed towards oneself" (Long 1986). The following passage from Diogenes Laertius is illustrative of the Stoic notion of *oikeiosis* (Long 1986, 185–86):

> The Stoics say that an animal has self-protection as the object of its primary impulse, since Nature from the beginning endears it to itself, as Chrysippus says in his first book *On goals:* "The first thing which is dear to every animal is its own constitution and awareness of this; for it was not likely that Nature gave it no attitude of estrangement or endearment. It follows then that having constituted the animal Nature endeared it to itself; thus, it is that the animal rejects what is harmful and pursues what is suitable (or akin) to itself."

The assertion that pleasure is the object of animals' primary impulse is proved to be false by the Stoics. For pleasure, they claim, if it really exists, is a secondary product when and only when Nature by itself has searched out and adopted the things which are suitable to the animals' constitution as such plea sure is like the flourishing of animals and the bloom of plants. Nature made no absolute distinction between plants and animals, for Nature directs plants too, independently of impulse and sensation, and in us certain processes of a vegetative kind take place. But since animals have the additional faculty of impulse, through the use of which they go in search of what is suitable to them, it is according to Nature for animals to be directed by impulse. And since reason in accord with a more perfect prescription has been bestowed on rational beings, life according to reason rightly becomes in accordance with their nature; for reason supervenes as the craftsman of impulse.

The following (premises 1–4) is an outline of the conceptual structure of the argument for *oikeiosis* that reflects the crucial features of that concept (Striker 1992):

(1) The first of the two impulses, the impulse for self-preservation, "fosters a self-regarding behavior such as looking after one's health and strength, getting food and shelter, acquiring material wealth, and also developing one's capacities for reasoning and trying to acquire knowledge – ways of acting that one might not ordinarily describe as being morally valuable."

(2) The second impulse, the impulse for sociability, was described by the Stoics as the foundation of justice, "based on respect for others and the desire to help fellow human beings seen as belonging to one, at least in kind." Following these two material impulses, the impulse for self-preservation and the impulse for sociability or caring for others, "would lead a human being to pursue and avoid things in a systematic, orderly way that can be described as "appropriate conduct."

(3) However, a person who led a "natural life" in this sense would not yet be considered virtuous. For virtue is not only a matter of acting correctly, but more importantly a matter of acting for the right reasons. According to the Stoics, virtue must be based on reason and knowledge, not on instinct. When eventually reason takes control over the natural impulses in a person's life, the person is: able to recognise that his natural and appropriate conduct exhibits a kind of rational order that is itself far more valuable than the objects pursued in his individual actions. He comes to realise that goodness consists of perfect rationality and thereupon makes it his ultimate aim to live according to the rational order of nature, of which human conduct is only a minor part. He will therefore cease to value natural advantages in their own right and continue to pursue them only in order to achieve agreement with nature and thus perfect his rationality. *The switch from the pursuit of natural aims to an exclusive interest in rational agreement with nature marks the transition to virtue* (emphasis added).

The "switch" in Gisela Striker's informed passage, marks a similar parallel "switch" in Gewirth's argument for the PGC. Namely, the crucial transition from self-regarding prudential rights to other-regarding moral rights mediated through the recognition that one has those rights on the sole basis of being a rational purposive agent.

(4) Finally, the virtuous person, the Stoic sage, "is the one whose every action is informed by his insight into the will of Zeus [or Nature] and whose only desire is to conform to nature's law."

The four stages described above comprise the central developmental stages in the Stoic doctrine of *oikeiosis*. Namely, the impulse for self-preservation; the impulse for sociability; and the recognition that one should not merely act appropriately towards oneself and others on the basis of "natural impulses" alone, but more importantly should do so on the basis of his rationality. And finally, the disposition and the practice of always acting according to one's perfected reason and virtue in agreement with nature, results in the attainment of eudaimonia. For this is in accordance with one's own rational nature that reflects and is part of, the rational order of nature, or the universe. Acting rationally is appropriate because it is in agreement with nature and agreement with nature is and ought to be the goal of life.

Neo-Stoicism of Alan Gewirth

Stages (1)–(3) of the Stoic doctrine of *oikeiosis*, as outlined above, are parallel to and schematically represent the developmental virtue of neo-Stoicism as I described it earlier. Stage (4), the final stage of sagehood, is parallel to and schematically represents the constitutive virtue of Gewirthian neo-Stoicism. All four stages of the Stoic doctrine of *oikeiosis* parallel and schematically represent in Stoic terms, the direct as well as the indirect applications of the argument for the PGC. Stage (4), the stage of sagehood, parallels and schematically represents in Stoic terms, Gewirth's model of ideal self-fulfilment and my ascription to him of the notion of a Gewirthian neo-Stoic sage; someone who makes the transition from developmental virtue to constitutive virtue and always acts and lives in accordance with perfect rationality and virtue on the basis of both the direct as well as the indirect applications of the PGC.

To be sure, Gewirth's argument for the PGC, both as regards its direct as well as its indirect applications, especially those concerning Gewirth's model of self-fulfilment, is not exactly the "cradle argument" of Stoic *oikeiosis*, for the earliest stage of childhood development is missing from the structure of the argument for the PGC. However, be this as it may, there is nothing in the formal structure of the PGC that cannot refer, retrospectively as it were, to this earlier stage of childhood development in Stoic *oikeiosis*. Granted that Gewirth's argument starts with mature rational agents (mature enough insofar as they are capable of minimally reasoning both inductively and deductively), there is no methodological reason why the argument could not, at least notionally, refer retrospectively to that earlier childhood development in Stoic *oikeiosis*. My comparison of Gewirth's argument for the PGC to the Stoic argument for *oikeiosis* takes up the comparison at a later stage of Stoic oikeiosis but still following the same pattern and rationale as the argument for oikeiosis. The following is my conceptual construction in outline of a parallel "Gewirthian neo-Stoic oikeiosis" (premises I–III):

Stage (I). Gewirth's argument for the PGC starts, as the Stoic argument of oikeiosis does, at the first stage of human development, namely, the impulse for self-preservation. It will be recalled that in the first stage of Gewirth's argument for the PGC, the agent is only concerned with his own particular purposes. It is his self-regarding concern for his own purposes that initially motivates the instrumentally rational agent to claim self-regarding prudential rights to his own freedom and well-being. This first stage of Gewirth's argument for the PGC parallels the first stage in the Stoic argument of oikeiosis. In both arguments, the agent's concern is primarily for himself and for his own desires, interests and purposes. The claiming of prudential rights by the agent in Gewirth's argument is motivated by self-interest which parallels, at least schematically if not in detail, the self-preservation impulse of the agent in the Stoic argument.

Stage (II). The second and crucial stage in Gewirth's argument for the PGC begins with the agent's realisation that he has prudential rights to his freedom and well-being just by virtue of being a purposive agent. It is this realisation that mediates the crucial "switch" in the agent's internal dialectical perspective from self-regarding prudential rights to other-regarding moral rights.

This stage in Gewirth's argument is parallel to Stages (2) and (3) in the Stoic argument for oikeiosis outlined above. Importantly, Gewirth's argument recognises that the impulse of sociability by being merely contingent is not sufficient or even necessary in providing the foundation for universal morality. What is required, as Stage (3) of the argument for oikeiosis brings out, is the recognition that virtue is not only a matter of acting correctly but more importantly acting for the right reasons; not merely a matter of acting out of a social instinct or impulse, that at best can only be contingent and variable, but a matter of acting rationally.

The Stoic switch from the pursuit of natural aims motivated by both the impulses for self-preservation and sociability, to an exclusive interest in rational agreement with nature that marks the crucial Stoic transition to virtue is, however, not clearly demonstrated by argument. We are led to understand that the belief that nature is rational and orderly together with the belief that one's own individual nature reflects and is part of that natural rational order, are together sufficient to rationally convince a per son that he ought to live in agreement with nature that, in turn, requires virtuous conduct. However, Gisela Striker correctly points out (Striker 1996) that the dual realisations that nature is rational and orderly and that one's own individual nature reflects and is part of that order, are not sufficiently compelling to support the transition from natural and social impulses to the rational "impulse" of acting and living virtuously in accordance with reason. For a person can ascent to the truth of both those propositions without being rationally compelled to accept the view that a good life is a life of virtue. For one can accept that nature is both rational and orderly and that one's own nature reflects that rational order without being rationally compelled to accept that that entails living virtuously and that living in agreement with nature is ethically good.

In Stage (II) of Gewirth's argument for the PGC closes that logical gap in the Stoic argument for oikeiosis by demonstrating that an agent is logically compelled on the basis of his own rational nature (rational purposive agency) to recognise and accept the transition from a self-regarding prudential perspective to an other-regarding moral perspective, on pain of self-contradiction (Gewirth 1978, 1996; Spence 2006). If my analysis is right, Gewirth's argument provides the Stoic argument for oikeiosis with the missing rational lacuna that closes the logical gap between the natural and social impulses in Stage (2) of that argument, and the transition to virtuous conduct, required by reason, in Stage (3).

In Stage (III), as mentioned earlier, the transition to constitutive virtue and Gewirthian neo-Stoic sagehood is provided by Gewirth's indirect applications of the PGC through his model of self-fulfilment. That final transition from developmental to constitutive virtue and ideal self-fulfilment parallels the Stoic transition from partial and imperfect virtue to complete and perfect virtue, a transition that marks Stoic sagehood. In conclusion, Gewirth's argument for the PGC mirrors, at least schematically if not in detail, the developmental structure of the Stoic argument for oikeiosis. In addition, the rational support provided indirectly for the argument for the PGC for both developmental and constitutive virtue as well as the arguments advanced by Gewirth on the basis of the argument for the PGC in support of his model of self-fulfilment, mirror and parallel the various Stoic arguments and doctrines for virtue and eudaimonia. Central to both Gewirth's argument for self-fulfilment and Stoic eudaimonia are the key elements of rationality, virtue and agreement with nature, which in Gewirthian terms translates to agreement with the logical implications of rational purposive agency and personhood.

The Core Problem of Stoic Philosophy and Technology

In both Stoic philosophy and technology, the key problem and question, and the focus of this final section of this chapter, and generally of this book, is the question of who is in control. Is it Us, as individuals and collectively as a society and the global community generally, or Technology? This final section will demonstrate that the answer provided by the Stoics to the question of who is in control of our lives, Us, or Circumstance, signals and motivates an answer to the question of who is in control of technology? Is it Us, as individuals and collectively as society, or the Big Tech companies who control that technology, and specifically, ICTs and AI technologies?

Who Is in Control? Us or Fortune?

Stoic philosophy sets out to answer the key meta-philosophical question of how to be happy in an unhappy world, despite the slings and arrows of outrageous fortune, a key feature Stoicism shares with Buddhism. The answer Stoicism provides is methodologically and logically ingenious, at once theoretical and practical. Perceiving philosophy to be a therapy of the mind, a Techne Viou or the Art of Living, the Stoics observed that there are things within our control and things outside our control. According to the Stoics, our main concern should be with developing our virtues, such as courage, moderation, justice, prudence or practical wisdom, empathy and solidarity, as virtue is the only inherent good, which provides both the necessary and sufficient practical means for the attainment of eudaimonia. To avoid unhappiness, we must do two things: control things that are within our control (our beliefs, judgements, desires, attitudes and volition or prohairesis) and be indifferent to things that are not in our control, or at least not in our full control, that is, things external to us, such as health, relationships, employment, wealth, power, death and so on. As rational beings, we should therefore perfect our characters through living a virtuous life in accordance with nature because it can bring us both ataraxia (tranquillity); autarkia (self-sufficiency) and eudaimonia (happiness or well-being). A happiness that is within our control, just because it depends on our own rational volition and judgements that are within our control. As we saw in Chap. 3, an important feature of Stoic eudaimonia is its intrinsic relation to virtue. For virtue as the only inherent good is not only the essential means for the attainment of eudaimonia but also constitutive of it. In being the ultimate goal of a good life, virtue is therefore both the means and the end for a eudaimonic or happy life, or at least, not an unhappy life. Eudaimonia, which in ancient Greek means having within us a good daemon or spirit, is up to us, as it lies within us and it is, therefore, within our control. Interestingly, the modern Greek word for happiness, *eftihia* (pronounced in modern Greek eftihia) that means having good fortune, is not within our control, as it places happiness outside our control, since fortune or circumstance, good or bad, lies outside our control.

The Core Problem of Technology: Who Is in Control? Us or Technology?

The core problem of technology, and specifically with regard to this chapter, ICTs, and in particular AI technologies, is that those technologies are not within our control but in the control of the Big Tech companies, such as Facebook and Google, as manifested in practice by the systemic surveillance of users' information by opaque and inscrutable algorithms, for the financial gain of Big Tech companies, dictated by their Business Model. That problem has been widely investigated by many researchers in recent years, notably Shoshanna Zuboff (2019) and a number of other researchers. The primary purpose of this chapter is to show how Stoic philosophy is both relevant and effective in theorising that problem, as well as practically responding to it for its solution. Having identified the parallel control problems of Stoic philosophy and technology, the chapter will show how Stoic philosophy is both theoretically relevant and practical to technology through its core principles we examined in detail in Chap. 3. To that end, this chapter will summarise in outline those core principles and show how they apply to technology, and in particular ICTs and AI technologies.

The Principles of Stoic Philosophy and Their Application to Technology

Principle 1: Concern Only for Things Within Our Control

According to Epictetus (2018), "Our opinions are up to us, and our impulses, desires, aversions – in short, whatever is our doing. Our bodies are not up to us, nor our possessions, our reputations, or our public offices, or, that is, whatever in not our doing ... So, remember, if you think that things naturally enslaved are free or that things not your own are your own, you will be thwarted, miserable, and upset, and will blame both gods and men" (Enchiridion, Ch.1:1–3) As the issue of Control is a core problem that arises in both Stoic philosophy and technology, Stoic philosophy is well placed to provide a theoretical and practical evaluation, as well as motivation to the solution of the control problem of technology. If core aspects of technology are outside our control, there are at least two options open to us within Stoic methodology:

> The first option is that since accuracy, reliability, trustworthiness, transparency, accountability, privacy, the inscrutability, monopoly and the asymmetry of information between Big Tech and us, as users of those technologies, and ultimately, the control of technology itself, is beyond our control, we should ignore such matters or at least treat them as preferred indifferents not required for our well-being, since we cannot control them.
>
> The second option is precisely because of the harmful impact those technologies have on the epistemic, ethical and eudaimonic features over most aspects of our lives, we should endeavour to bring those technologies under society's control, since our collective well-being is at risk if we don't.

A key underlying assumption of the Stoic notion of control is inner mental freedom and autonomous rational agency. However, the psychologically per suasive technologies used by the Big Tech companies to inculcate users to spend maximum time on Big Tech's digital platforms (since more time spent is more money earned by them) are designed to undermine the users' rational autonomous agency for the primary financial gain of those Big Tech companies. Which, like concealed and inscrutable puppeteers want to make us dance to their Faustian-Midas' tune, just as the prisoners in Plato's Allegory of the Cave. In Chaps. 6, 7, 8, and 9, we shall examine more closely how Big Tech companies, such as Facebook and Google, using their surveillance Business Models through the application of opaque persuasive algorithms, undermine the users' rational autonomous agency for the primary financial gain of those companies: by collecting and selling users' data to advertisers and other information contractors, without the users' explicit knowledge and consent.

Principle 2: The Final Goal

The Hellenistic philosophers, including the Stoics, claimed that eudaimonia or happiness was the ultimate goal of a good life, though each of the Hellenistic schools identified happiness with something different: The Epicureans, identified it with pleasure and the avoidance of pain, the Pyrrhonian Sceptics with suspension of all judgement as the only sure way to achieve ataraxia or tranquillity, and the Stoics identified eudaimonia exclusively with a life of virtue, which, in turn, they identified with a life in agreement with nature. It is, so to speak, our "nature" to be rational, and it is our rational nature that leads us through various intermediary stages, to a virtuous and ethical life that ultimately results in the attainment of eudaimonia or happiness.

The emphasis of the Stoics on the attainment of eudaimonia through virtue is significant for examining the impact of technology on our individual and collective well-being, and how we can respond to it. It highlights why technology cannot be good for itself or for the Big Tech companies alone that manage it, but for the common good of all humanity, which was the topic we examined in Chap. 2.

Principle 3: The Life of Virtue and the Life of Wisdom

As we saw earlier, the Stoic definition of virtue is as follows:

(1) Virtue is not defined by the consequences in the world which it succeeds in promoting, but by a pattern of behaviour that follows necessarily from a disposition perfectly in tune with Nature's rationality.

(2) The right thing to do is that which accords with virtue, and this is equivalent to saying that it accords with the nature of a perfect rational being.
(3) Virtue accords with Nature in the sense that it is the special function or goal of a rational being to be virtuous. (Long 1986, 192)

Associated with the Stoic principle of a virtuous life is the principle of Wisdom. The ultimate object of philosophy according to the Stoics, and the ancient Greek philosophers generally, beginning with Socrates, is to teach us not just knowledge, but wisdom. Wisdom, therefore, for the Stoics is a way of a life of virtue that brings peace of mind, inner freedom, happiness and a cosmic consciousness. By "cosmic consciousness," the Stoics understood that quality, as universal reason, which permeates the whole universe and by virtue of which we are all, integrated parts of the cosmos. As Pierre Hadot tells us (1995),

> The exercise of wisdom entails a cosmic dimension. Whereas the average per son has lost touch with the world, and does not see the world qua world, but rather treats the world as a means of satisfying his desires, the [Stoic] sage never ceases to have the whole constantly present to mind. He thinks and acts within a cosmic perspective. He has a feeling of belonging to a whole, which goes beyond the limits of his individuality.

Significantly, more than information, wisdom requires transformation. To become wise, one requires not only to be informed but to become transformed, through living virtuously, by the practice of philosophy as a way of life. Given the unwise ways in which we use technology, including our uncritical acceptance of the control Big Tech companies exert over all facets of our lives, we should use information, more virtuously and wisely for the collective good and well-being of ourselves and society. Wisdom, as meta technology of the mind (Spence 2011) can guide us on how to control technology and our use of it, for our common good. The topic of wisdom has been explored in more detail in Chap. 2.

Principle 4: Cosmopolitanism

> When anyone asked him where he came from, he said, "I am a citizen of the world." (Laertius 1925, Chapter 2, 22–84)

Closely associated with the notion of cosmic consciousness is the Stoic cosmopolitan perspective. It involves living empathetically and in solidarity for the common good of the world community and acting in accordance with global social justice for our collective well-being. According to the Stoics, philosophy entails a community engagement. For the Stoics, however, the polis was the cosmopolis, not the city state, or individual nations, but the whole world (Spence 2007).

The threat to the natural environment through climate change and the recent COVID-19 pandemic renders the ethical lifestyle recommended by the Stoics, which respects the ethical rights of all individuals as citizens of the world, and also respects both the natural and digital environments, not only desirable, but necessary. In the connected technological age in which we live, Stoic cosmopolitanism as a way of life for the attainment of well-being is relevant and important for enabling us to mount a collective response to Big Tech's surveillance practices that systemically undermine our collective well-being. Since we cannot mount such a response as individuals, our response, just as our response to a solution to climate change, requires a global collective action that incorporates citizens, governments and their associated democratic institutions and regulatory authorities, as well as the Big Tech companies themselves. Stoic cosmopolitanism provides both the theoretical framework and practical motivation for such a collective shared response.

Principle 5: A Life in Agreement with Nature

> *As Cicero explains, the value of anything in Stoicism is defined by reference to Nature.* (Long 1986)

An important and core principle of Stoicism is living in accordance with nature. Eudaimonia, according to the Stoics, consists in living a virtuous life and a virtuous life is one in agreement with nature, which they identified with reason or rational spirit that permeates the entire universe. Since nature was identified with universal reason, the Stoics identified a life in agreement with nature with the perfection of human reason. Moreover, it is only by living a rational and virtuous life that human beings can truly become happy. For the Stoics, there is an integral relationship between virtue, rational agency, nature and well-being, all of which relate to living a good life for the attainment of eudaimonia.

Since, the Stoic notion of nature, under the influence of Heraclitus, was conceived to be dynamic and not static, we can extend the Stoic notion of nature to include not only the natural environment but also crucially the digital environment. Applying the Stoic principle of living in agreement with nature we can say that just as we have a collective global responsibility in respecting and safeguarding the integrity of the natural environment (biosphere), we have an equal responsibility with regard to the digital environment (infosphere). Whereas climate change pollutes and corrupts the biosphere, the surveillance practices of the Big Tech corrupt the infosphere. Both are a form of corruption that undermine both our individual and collective global well-being. In Chap. 6, we shall examine the notion of media corruption with regard to some of the information and communication practices of Big Tech companies, such as Facebook and Google.

Principle 6: Oikeiosis

Related to the issue of transformation through wisdom examined earlier, is the principle of oikeiosis. As we saw in Chap. 3, the transition from developmental to constitutive virtue, from imperfect and incomplete virtue to perfect and complete virtue, requires a kind of ethical transformation. In an insightful passage, Julia Annas describes quite clearly the kind of ethical transformation that the Socratic theory of happiness requires and which by extension applies equally to Stoic philosophy, as the Stoics considered Socrates to be their earlier role model (Annas 1999):

> As a theory of happiness [the Socratic theory of happiness in Plato], it is very demanding. You won't have got your account of happiness right until you achieve the right perspective, and that will be utterly different from your initial standpoint. Becoming virtuous requires transforming [emphasis added] your values and priorities; nothing short of this will do. All your valuations of conventional goods and evils will turn out to be wrong … we are used to moral theories that are demanding, but it is not easy for us to conceive of such a theory which is a theory of happiness, rather than something else. But this is not an argument against a theory like Plato's. It shows merely that it takes a long intellectual journey away from common sense and conventional beliefs to achieve the wisdom of the virtuous person and thus to have your whole life, and your conception of happiness transformed.

Alan Gewirth also recognises and acknowledges the transformative aspect of his concept of ideal self-fulfilment. As he tells us (Gewirth 1998),

> [T]o recognize and accept these requirements in one's thought, and action may involve a kind of self-transformation [emphasis added] whereby one becomes more fully a moral and social being.

It is the realisation of the practical necessity for such transformation and the need for providing a conceptual account of how such a transformation can practically take place, that the Stoics developed their theory of oikeiosis; a theory considered to be one of the most crucial and fundamental doctrines of Stoic ethics. The theory purports to show how a person can progressively develop, because he is naturally conditioned and thus capable of doing so, from the state of having and being motivated by purely hedonistic desires as a child, to the state of acquiring, over a time period of biological, psychological, social, but more importantly, rational development, an ethical perspective.

As we saw earlier, the four stages comprising the central developmental stages in the Stoic doctrine of oikeiosis are: the impulse for self-preservation; the impulse for sociability and the recognition that one should not merely act appropriately towards oneself and others on the basis of "natural impulses" alone, but more importantly should do so on the basis of their rationality. And finally, the disposition and the practice of always acting according to one's perfected reason and virtue in agreement with nature, results in the attainment of eudaimonia. For this is in accordance with one's own rational nature that reflects and is part of the rational order of nature and the universe. Acting rationally is appropriate because it is in agreement with nature and agreement with nature is necessary if one is to attain eudaimonia, which is the ultimate goal of a good life (see Principle "The Final Goal" above). As we saw in Chap. 2, the DOIT-Wisdom which advances a neo-Stoic philosophy based on a reconstruction of Gewirth's theory of self fulfilment, examined above, aligns with the Stoic notion of oikeiosis. Gewirth's argument for the principle of generic consistency (PGC), which supports universal human rights for freedom and well-being, mirrors, at least schematically if not in detail, the developmental structure of the Stoic argument for oikeiosis.

In addition, the rational support provided indirectly for the argument for the PGC for both developmental and constitutive virtue as well as the arguments advanced by Gewirth on the basis of the argument for the PGC in support of his model of self-fulfilment, mirror and parallel the Stoic arguments and doctrines for virtue and eudaimonia. Central to both Gewirth's argument for self-fulfilment, and Stoic eudaimonia are the key elements of rationality, virtue and agreement with nature, which in Gewirthian terms translates to agreement with the logical implications of rational purposive agency and personhood. And as we saw earlier in examining the problem of control regarding technology, the problem is how Big Tech companies, such as Google and Facebook, target our rational autonomy and agency through their systemic persuasive opaque algorithms for their own self-centred financial gain, and do so without any transparency and accountability.

Applying the principle of oikeiosis to technology further emphasises the urgent need why we must transform and no longer accept what is clearly harmful to us. Namely, the wilful deception and manipulation of our autonomy and privacy by Big Tech companies. In later chapters of this book, and especially in Chaps. 7 and 8, we shall examine in more detail how the concepts of autonomy and privacy are closely interrelated and both essential for our fundamental rights to freedom and well-being as well as our dignity, as individuals and collectively as society.

In the introduction of her important book, *Privacy is Power,* Carissa Véliz (2020, 1) sounds a siren warning of how Big Tech violates and usurps our privacy, thus undermining our individual and collective well-being:

> They are watching us. They know I'm writing these words. They know you are reading them. Governments and hundreds of corporations are spying on you and me, and everyone we know. Every minute of every day. They rack and record all they can: our location, our communications, our internet searches, our biometric information, our social relations, our purchases, and much more. They want to know who we are, what we think, where we hurt. They want to predict and influence our behaviour. They have too much power. Their power stems from us, from you, from your data. It's time to take back control [emphasis added]. Reclaiming privacy is the only way we can regain control of our lives and our societies.

Taking back control of technology by taking control of our autonomy and privacy, as individuals and collectively as society, is also the intended clarion call of this book, guided by the normative and eudaimonic conceptual framework and rational of Stoic and neo-Stoic philosophy. For as this chapter has shown the solution provided by Stoic and neo-Stoic philosophy to the control problem of who is in control of our lives, Us or Circumstance, offers also a response to the control problem of technology, that is, who is in control of technology, Us or Big Tech. It is a matter of necessity, therefore, to mount a collective cosmopolitan response to Big Tech, as the condition for our well-being and ultimately our survival as free and autonomous human beings, depends on it. As we shall see in Chap. 8, when examining in more detail the control problem of AI technology, the survival of humanity itself, depends on it.

Principle 7: Philosophy as a Way of Life

Empty is that philosopher's argument by which no human suffering is therapeutically treated. For just as there is no use in a medical art that does not cast out the sicknesses of the body, so too there is no use in philosophy, unless it casts out the suffering of the soul. (Epicurus, quoted in Nussbaum 1994)

An important aspect of Stoic philosophy, as well as the other Schools of Hellenistic philosophy, and in particular, Epicurean philosophy as the quote from Epicurus above indicates, was that philosophy had to be conducted as a way of life or the art of living. Philosophy, as we saw in Chap. 3, can be viewed as the art of living in two ways:

1) As a craft or art which once learned can enable one to live a good and fulfilling life on a daily basis and
2) The perfection of one's human nature—becoming the best of one's kind—humankind. For the Stoics, the perfection of one's human nature meant the perfection of one's rational nature. For according to the Stoics, the essence of human nature and nature generally, the whole universe in fact, was reason. Thus, in perfecting one's rational nature one became at once the object and the subject of art, that is, the art of life. This was the highest form of art, in perfecting one's rational nature through the practice of philosophy. In the neo-Stoic theory, I based on my reconstruction of Gewirth's notion of self-fulfilment, a life in agreement with nature is a life in agreement with the perfection of one's rational purposive agency, which, in turn, is in agreement with universal morality, which is essential for a good and self-fulfilled life. In contrast to the Stoic conception of universal nature as divine, however, Gewirth's notion of "nature" is and idealised human nature, understood as rational purposive agency applied to the aim of becoming the best oneself as a human being,

which as we saw earlier requires nothing less than a transformation through the exercise of wisdom. Wisdom, as a form of experiential meta-knowledge of how to live a life of virtue for the attainment of eudaimonia, requires philosophy to be lived in practice as a way of life and not merely an intellectual exercise of thought alone. As Pierre Hadot tells us (1995),

> Philosophy as a way of life for the Stoics was, A mode of existing-in-the-world, which had to be practised at each instant, and the goal of which was to transform the whole of the individual's life ... the word philosophia—the love of wisdom—was enough to express this conception of philosophy.

Significantly, the application of Stoic and neo-Stoic philosophy as a way of life in addressing and responding to the problem of technology examined above, namely the problem of control by the Big Tech companies over all aspects of our lives, is well suited to a global digital network environment: both for its identification and eudaimonic evaluation of its impact on society, as well as the enablement of its solution, through the shared collective normative response by users of technology and society and the global community generally. For just as Big Tech can use the user's information to the detriment of their users' well-being, the users themselves have the capacity and ability to apply their shared knowledge of the control problem of technology to turn the tables on Big Tech. That is because the users can at a stroke put a significant dent in the Big Tech's business model of clickbait and consequently their billion-dollar revenue stream, obliging the Big Tech companies to take appropriate action to remedy their dis-demonic manipulative practices against the users' legitimate normative claims.

For just imagine for a moment what would happen if the billions of Facebook's and Google users worldwide went on strike and stopped using their platform services, even for a day. Both companies would lose a lot of money, and their surveillance-based business model would shudder. Of course, the practicality of such shared collective action may prove difficult but as they say, where there is a will, there is a way. Stoic philosophy is at least theoretically and practically well equipped to motivate and provide both the will and the way.

Conclusion

In the connected digital age, Stoic philosophy as a systemic unified practical philosophy of human flourishing and well-being, which combines epistemic, ethical and eudaimonic principles clustered around the central concept of wisdom, is eminently relevant and able to respond and motivate both theoretically and practically a solution to the control problem of technology.

In keeping with the Stoic principles of living virtuously, rationally, in accordance with nature, for our individual and collective well-being, as cosmopolitan citizens with a comic consciousness for the attainment of well-being, we have a collective responsibility, therefore, to respond to the problem of Big Tech's monopolistic control over the digital environment that has an overall dis-demonic impact on our individual and collective lives.

As a secular cosmopolitan philosophy, with its naturalistic and public outlook and methodology, as a way of life, Stoicism has for some time now been experiencing a renaissance both in the academy and recently in the public domain through a strong worldwide online presence. The DOIT-Wisdom model we examined in Chap. 2, further illustrates the application of Stoic and neo-Stoic philosophy to technology. The Stoic cosmopolitan global community, therefore, has the capacity and the ability to generate and motivate a worldwide public response to the control problem of surveillance by the Big Tech companies. As a long-time fan of Star Trek, let me add that Resistance is not Futile.

As my friend and colleague Rick Benitez, at the University of Sydney told me recently when the pandemic lockdown was first introduced, "Stoic Philosophy is the only game in town." I will not dispute that rhetorical flourish, though I will add that Aristotelian and Epicurean philosophy, as well as Buddhism and Hindu philosophy, as well as some indigenous philosophies, with which Stoicism has some suggestive similarities in regard to nature, are not excluded by Stoic and neo-Stoic philosophy

References

Annas, Julia. (1999). *Platonic Ethics, Old and New*. Ithaca: Cornell University Press.
———. (1993). *The Morality of Happiness*. New York: Oxford University Press.
Becker, Laurence. (1998). *A New Stoicism. Princeton: Princeton University Press.*
Cicero, Tusculan Disputations (Loeb Classical Library) 2nd Edition, translated by J. E. King. Cambridge, MA, and London: Harvard University Press.

Epictetus. (2018). *Epictetus: Discourses and Selected Writings*, translated and edited by Robert Dobbin. London: Penguin Classics

Friedman, Marilyn. (2000). "Educating for World Citizenship," *Ethics*, April, 589.

Gewirth, Alan. (1998). *Self-fulfillment*. Princeton: Princeton University Press.

———. (1996). *The Community of Rights*. Chicago: University of Chicago Press.

———. (1978). *Reason and Morality*. Chicago: University of Chicago Press.

Hadot, Pierre.(2001). *The Inner Citadel: The Meditations of Marcus Aurelius*. Cambridge, MA: Harvard University Press.

———. (1995). *Philosophy as a Way of Life*. Oxford: Blackwell Publishing, 273.

Laertius, Diogenes. (1925). *Lives of Eminent Philosophers Books 6–10*. Cambridge, MA: Harvard University Press.

Long, A. A. (1986), *Hellenistic Philosophy – Stoics, Epicureans, Sceptics*. 2nd Edition. Berkeley: University of California Press.

Nussbaum, Martha. (2000). "Duties of Justice, Duties of Material Aid: Cicero's Problematic Legacy." In *Symposium on Cosmopolitanism. The Journal of Political Philosophy*, 8 (2):191.

———. (1997). *Cultivating Humanity: A Classical Defense of Reform in Liberal Education*. Cambridge, MA, and London: Harvard University Press.

———. (1994). *The Therapy of Desire*. Princeton, NJ: Princeton University Press.

Spence, Edward H. (2011). "The Theatre of Philosophy: A Neo-Socratic Method for Performing Public Philosophy", In Bull, Gordon (Ed.), *Creativity: Brain – Mind – Body: A View into the Future of Australian Art and Design Schools*, the 2011 Australian Council of University Art and Design Schools (ACUADS) Conference, the Australian National University, the University of Canberra, and the Canberra Institute of Technology, 21–23 September.

———. (2007). "Positive Rights and the Cosmopolitan Community: Right-Centered Foundations for Global Ethics." *Journal of Global Ethics*, 3 (2): July/ August New York: Taylor and Francis.

———. (2006). *Ethics Within Reason: A Neo-Gewithian Approach*. Lanham, Maryland: Lexington Books, a division of Rowman and Littlefield, 433–442.

———. (2004)."Philosophy Plays: A Neo-Stoic Method for Teaching Ethics." *Teaching Ethics*, 5 (1):41–57.

———. (1999) "A Commentary on Deryck Beyleveld's "Gewirth and Kant on Justifying the Supreme Principle of Morality." In Michael Boylan (ed.), *Gewirth, Critical Essays On Action, Rationality, and Community*. Maryland: Rowman and Littlefield Publishers, Inc. pp 119–124. ISBN: 9780847692590.

Striker, Gisela. (1996). *Essays on Hellenistic Epistemology and Ethics*. Cambridge: Cambridge University Press

———. 1992. "Stoicism" In Lawrence C. Becker and Charlotte B. Becker (Eds.), *Encyclopedia of Ethics*. New York: Garland Publishing, 1208–1213.

Véliz, Carissa. (2020). *Privacy is Power: Why and How You Should Take Back Control of Your Data*. London:Bantam Press.
Zuboff, Shoshana. (2019). *The Age of Surveillance Capitalism: The Fight for a Human Future at the New Frontier of Power.* London: Profile Books.

CHAPTER 4

Artificial Agency and Moral Autonomy: The Quest for Artificial Wisdom

Come writers and critics
Who prophesize with your pen
And keep your eyes wide
The chance won't come again
And don't speak too soon
For the wheel's still in spin
And there's no tellin' who
That it's namin'
For the loser now
Will be later to win
For the times they are a-changin'.
—Bob Dylan (1964)

Introduction

This chapter will address a very important topic in the Philosophy and Ethics of Technology, which is that of *artificial mediated agency and autonomy* and its moral significance and impact on human well-being. Furthermore, it will address the key question of this chapter whether artificial agency and autonomy can be *extended to the notion of wisdom*: if agency is mediated and distributed between humans and intelligent machines, can there also be a mediated wisdom? Although several well-known philosophers of technology, such as Luciano Floridi (2013, 2014),

E. H. Spence, *Wisdom in the Age of Intelligent Machines*,
https://doi.org/10.1007/978-3-031-93883-2_4

Peter-Paul Verbeek (2011, 2014), Richard Heersmink (2017), to mention but a few, have researched and written on the issue of ***artificial moral mediation***, very little research has been published on the topic of wisdom with regard to AI technology and its impact on well-being. This chapter and the book in general, sets out to address and fill that gap. According to Verbeek, "only by recognizing this interweaving of human and technologies can we take responsibility for the ways in which ***technologies have an impact on society and on human existence*** [my emphasis] – in practices of technology design, implementation, and use" (2014). In a clear summary Heersmink summarises Verbeek's distributed moral agency (2017) thus:

> By drawing on actor-network theory, Verbeek (2011) develops the notion of "distributed moral agency." He specifically focuses on the moral aspects of obstetric ultrasound imaging, an imaging technology used to visualize a foetus in the mother's womb, which is usually done a week twelve and twenty of pregnancy. This technology does not just provide a neutral and transparent window to the womb but has several morally important aspects...so obstetric ultrasound imaging mediates and transforms the relationship parents have with their foetus, providing them with information on the health of their foetus that is important for moral decision-making regarding pregnancy and abortion. For these reasons, moral agency should be seen as distributed across humans and technology. In Verbeek's words: Ultrasound imaging actively contributes to the coming about of moral actions and the considerations behind moral actions. The example therefore shows that moral agency should not be seen as an exclusive human property; it is distributed among human beings and nonhuman entities. (2011)

In a later chapter he points out: Moral mediation always involves an intricate relation between humans and nonhumans, and the "moral agency" that results from this relation therefore always has a hybrid rather than a "purely human" character.... moral agency is distributed among humans and nonhumans; moral actions and decisions are the product of human-technology associations (2011).

According to Heersmink, "Verbeek ... argues that moral agency is distributed across humans and technology, they cannot be understood in isolation from each other because they are integrally connected. So, one way to understand the moral relevance of technological artefacts is to say that artifacts are co-constitutive of moral agency" (2017).

However, Verbeek, Floridi and Heersmink and the other aforementioned philosophers, do not refer to the important role that wisdom should play in recognising and assessing the mediated impact of technology on well-being, and it is precisely that missing inclusion of wisdom and its gap that this chapter and this book overall, intends to fill. Namely, to examine and evaluate the impact of mediated AI technologies on our well-being, in terms of our agency, autonomy, morality and most importantly, as we shall examine in Chaps. 6 and 7 of this book, our dignity, and do so through the concept of wisdom. One key question this chapter will examine is whether there can be mediated wisdom along the lines that Floridi, Verbeek and Heersmink apply to mediated moral agency or is some other way that retains the important role of human autonomy and dignity (Ardelt & Edwards, 2016), which is of central importance to human agency, a topic that is examined in Chap. 7 of this book. In an important innovative chapter "Artificial wisdom: a philosophical framework" Cheng-hung Tsai examines the notion of artificial wisdom (2020). In his chapter he explains "why artificial wisdom maters and how artificial wisdom is possible (in principle and in practice) .... the result is a conceptual framework that guides future research on creating artificial wisdom" (2020). Contrary to that view, Jason Swartwood and Valerie Tiberius (2019) argue:

> A wise person's understanding is *uncodifiable:* it is possible to derive good guidance in any particular situation (Hursthouse 1999: 39–40). Morality is too complex, Virtue Ethicists argue, to capture with such a set of principles. For instance, someone might try to defend the following moral principle: An action is right when, of all the actions possible in a situation, it could be expected to produce the most total long-term happiness when the effects on everyone are added up. If this principle accurately described what matters in all situations, then a non-virtuous person could use it to derive (after suitable inquiry into the likely consequences of different actions) reliable guidance about what to do. Virtue Ethicists argue that principles like this do not succeed in codifying wise understanding.

Julia Annas provides additional arguments against codifiability (Annas 2004):

> If Virtue Ethicists are correct that wise understanding is uncodifiable, then that would imply that a reliable understanding of how to conduct oneself must be acquired through some sort of experience and reflection.... Aristotle thought we need reflection that focuses on both the universal and the particulars of situations. (NE 1141b10-15, 1142a14)

Swartwood and Tiberius raise an important characteristic of wisdom that resists codifiability, that is, wisdom is a key characteristic of Virtue Ethics and by extension Stoic philosophy and Hellenistic philosophy generally, which is that wisdom is a way of life for the attainment of eudaimonia as argued in Chap. 3 of this book, and thus specific to individual lives. This chapter will examine in further detail whether even if uncodifiable, wisdom, however, could be mediated in some way to be defined, through AI technology as suggested with regard to morality by Verbeek and Floridi, whilst at the same time, preserving the moral agency of wise human beings that remains under their full control.

Thoughts and Theories of Artificial Wisdom

Artificial Wisdom: A Philosophical Framework

As mentioned in an important innovative chapter referred to in the Introduction above, Artificial wisdom: a philosophical framework" Cheng-hung Tsai examines the notion of artificial wisdom (2020). In his chapter he explains "why artificial wisdom maters and how artificial wisdom is possible (in principle and in practice) .... the result is a conceptual framework that guides future research on creating artificial wisdom" (2020). As he proceeds to say, "if AI researchers aim to build machines to mimic humans with almost every aspect of human excellence, then practical wisdom is likely to be of the utmost interest."

As argued in Chap. 2 of this book with regard to my tripartite notion of wisdom, to answer that question, whether AI agents have the theoretical and practical capacity to acquire a notion of Artificial Wisdom (AW) on the basis of a theoretical groundwork, may provide such an answer; As the concept of *wisdom* understood as a type of *meta-knowledge* as well as a type of *meta-virtue*, which can enable us as individuals and collectively as a society, and humanity generally, to both know in principle what constitutes a good life, and how to successfully apply that knowledge as *know-how* in living such a life in practice, for the ultimate purpose of attaining well-being or eudaimonia, could in principle be applied to Artificial Wisdom. This answer will be based on the main argument presented in this chapter that the notion of *wisdom* understood as a triadic concept, at once a *meta-epistemological*, *meta-ethical* and *meta-eudemonic*, provides at least theoretically the essential conceptual link between information on the one hand and well-being on the other. In the digital age of information, both the theoretical examination and analysis of the question of how information in its technological construction and dissemination relates to well-being, and the provision of an adequate answer to that question are essential for developing a deeper understanding of how to evaluate the theoretical and practical implications and impact of digital information on well-being for individuals and societies living with AI agents.

In sum, the chapter seeks to show that although related, wisdom is conceptually distinct and conceptually different from information, intelligence and knowledge in a crucial way. For wisdom unlike information and first-order knowledge provides a person with *understanding* concerning the *techne viou* or *craftsmanship of living* in the sense of *knowing how* to *evaluate and apply* relevant information or knowledge for living a good life for the attainment of eudaimonia, and in addition, an appreciation in knowing *why* such a life constitutes a good life. For the Stoics, and as we saw in Chap. 3, and generally for the Hellenistic philosophers, philosophy was conceived as a way of life or *techne viou*. As such, Wisdom as a triadic model comprising epistemic, ethical and eudaimonic components, can be thought of as a Way of Life, as this chapter and the book on the whole will demonstrate, and as such can be applied to evaluate the impact of ICT and AI technologies on our well-being both individually and collectively as a global community, not only human but also digital, where AI technology is an important aspect.

As Cheng-hung Tsai astutely observes, in contrast to instrumental, mans-end reasoning where the reasoning is about the given goal, "wisdom regardless of whether it is human or artificial, should go beyond mere mans-end reasoning; an agent with practical wisdom is able to deliberate well about the final goals. Because intelligence lacks what is crucial to being wise." And what is crucial, is the knowledge that leads to the acquisition of well-being or *eudaimonia*. (2020).

Beyond Artificial Intelligence: Exploring Artificial Wisdom

In another innovative paper, D.V. Jest et al. argue:

It is not intelligence, but wisdom, that is associated with greater well-being, happiness, health and perhaps even longevity of the individual and the society (Jeste et al. 2019). Thus, the future need in technology is for

AW and this will also serve to mitigate the risks of advanced AIs such as Superintelligence. Human wisdom is a scientific construct supported by empirical research during the last 45 years. It has a scientifically based definition, validated measurement scales, underlying neurobiology and genetics, and relationship to aging that are all quite different from those for human intelligence. AW would be modelled after human wisdom and contribute to our understanding of wisdom by clarifying mechanisms or psychological underpinnings. AW development will require close collaboration among computer scientists and engineers, neuroscientists, mental health experts and ethicists to ensure that AW is developed in a way that best complements and supports human lives.

As we examined in Chap. 2, not only the pursuit of knowledge alone, but also that of morality (the virtues) and well-being (eudaimonia) are essential components of wisdom. Thus, any notion of AW will have to show how AI agents can acquire and exercise those three components.

D.V. Jeste et al. go on to further explore the question whether AW must possess "artificial consciousness" citing (Buttazzo 2001) arguing claiming only humans can have consciousness and that only humans can truly be wise. I tend to disagree with them as the question of AI consciousness is still very much an open question (Susan Scheider 2019) and one not settled yet with any degree of certainly. As those authors argue with regard to AW humans can develop AI systems to perform actions that use and apply wise principles that result in wise acts (Sevilla 2013). The key question as they correctly point out is whether computers can be designed to adapt and become "wiser" in their algorithms.

According to these authors, the three principles they identified for AW systems are:

(1) A wise system will learn from experience (self-reflection).
(2) A wise system will exhibit pro-social behaviours (empathy, compassion) in interacting with individuals and model social decision-making (social advising).
(3) A wise system will be able to discern human emotions and help people to emotionally regulate and exercise good judgement or knowledge.

Thus, AW can help promote emotional regulation in it users, according to those authors. They go on to say that a wise system could act as a "wisdom coach" "to provide a cognitive reappraisal of a stressful situation through appropriate role modelling and thus help the user to reinterpret the meaning of a distressing event and control negative emotions." Which as we saw earlier in Chap. 3. Stoic philosophy is what recommends which partly explains the worldwide popularity of Stoic philosophy today.

An important point the authors make that is worth mentioning here, is that "a wise system would need to promote personalised wisdom for different users under different condition and at different times." As the authors go on to conclude, "intelligence is necessary but not sufficient for wisdom," which was what I emphasised when describing my tripartite notion of wisdom in Chap. 2. What those authors go on to emphasise and I fully support, is that "AI technologies of the future will require new conceptual and computational models based on human wisdom and not human intelligence, to produce AW." An AW agent can serve as an as an assistant, peer, or coach, to help make humans wiser. The authors conclude:

> A long-term transdisciplinary collaboration is *essential* (emphasis added) for the development of AW, which will have a far greater impact positive impact on human well-being than any technology developed to date. (2019)

My own notion of this development that I fully support and promote, is a Hybris Notion of Wisdom, one shared between Human Wisdom and Artificial Wisdom.

Can AI Systems Become Wise?

According to Ana Sinha and Pooja Lakhanpal from the Indiana Institute of Foreign Trade, New Delhi, India (4 May 2024) "although we do not have a common definition of wisdom, nor do we have an established methodological protocol to analyse or measure it. The reason for this they argue this deficiency is the issue of subjectivity and bias." However, they claim the AI systems in their generative forms have swept these elements of subjectivity, biases and numbers away on the basis of the sheer volume data, on which their algorithms work to make choices. According to them a generative AI tool like ChatGPT that has access to the global search engines and data in its voluminous capacities, the decision records on which they made their decisions should be considered wise. For the artificial wisdom (AW) concept is built on the notion of evolving AI systems into wise systems that comprises two parts, the datasets and the generative, make their predictive modelling building on these data sets, independent from human intervention. A good example of that is when AlphaGo of Google's Deep Mind beat the world champion of Go, Lee Se-dol, considered the greatest Go player, whose his defeat by AlphaGo software was seen as landmark moment for artificial intelligence. As the two authors claim, the wisdom question is in many ways at the heart the AI debate and raise the question, can AI systems become human like? Can they as predicted exceed humans not only in knowledge, which they have already done but also in wisdom? Can they be as wise as Socrates?

Another important issue the two authors raise is that the development and deployment of AI systems are influenced by the interests of those who control the technology, the Big Tech companies such as Google, Facebook, Amazon, Apple and Microsoft, an issue raised in Chap. 7 of this book, namely, he Control Problem of AI. As the two authors correctly ask, "who is responsible for the decisions made by AI systems?" We can ask the further question are they wise? For can wisdom be controlled by the few billionaires ruling over the many. Another important question the pose is that of the alignment problem, is whether the decisions made by AI systems align with human values and social norms? As we shall see in Chap. 7 of this book, the alignment problem is at present unsolvable. If that is still the case how can Human Wisdom be aligned with Artificial Wisdom? Perhaps as the two authors and those we examined above previously, the generative AI systems and there speedy rise in the last two years and beyond, heralds, the obvious question of artificial wisdom. For as the two authors above point out, "the AI systems continue to play an increasingly important role in our lives, it is *critical* [emphasis added] that we work towards ensuring that they make wise and ethics decisions that align with human values and societal norms and importantly contribute to our collective wellbeing, globally."

From Artificial Intelligence to Artificial Wisdom: What Socrates Teaches Us

In an invigorating paper Tae Wan Kim and Santiago Mejia (2019) examine and argue for a Socratic model for Artificial Wisdom. Beginning with the question "what is an excellent, flourishing, human being," they go on to say: "Socrates taught us two important things about this question: 1) reflecting on it was a central part of being human and 2) engaging with this question leads to the recognition of a particular form of ignorance that is also a form of wisdom." This method of doubt was of course the method of *Socratic Elenchus* that involves the assumption of ignorance which introduces *aporia* (doubt leading to curiosity to learn) which, in turn, leads to the acquisition of knowledge and then wisdom: knowledge of what is good or evil. As the authors point out, "what Socrates claimed as ignorance was a form of wisdom." For Socratic wisdom, via ignorance, brings openness and humility with respect to how the most important human questions should be answered. Socrates wanted to become wiser by conversing with anyone about human flourishing, regardless of class, social status and geographical origin for the thoughts of himself as a cosmopolitan, a citizen of the world. The fact that he was willing to examine everyone and was open to all opinions makes his approach a powerful tonic against "echo chambers and filter bubbles" in our digital information age. An important point they raise is that AI currently cannot do causal/counterfactual reasoning. They argue that a fundamental premise of this argument is that an important feature of human intelligence is causal/counterfactual thinking and would be good if AI would replicate human intelligence. However, the two authors add that apart from counterfactual reasoning AI should also incorporate the kind wisdom that the oracle attributed to Socrates, wisdom as a form of ignorance and humility on which the Socratic Elenchus, referred to earlier, is based. They conclude:

> Socratic Siri would not simply aim to deliver information matter of factly. It would help users be more reflective by challenging them to critically engage with the material they are consuming and ensuring that such consumption is connected with the reflection of what it means to live a good life.

I will add the life based on virtue and wisdom that leads to the well-being and that of their fellow human beings worldwide, in accordance with the cosmopolitan view of the Stoics.

Smarter AI Assistants Could Make It Harder to Stay Human (Steven Levy 2023)

According to Steven Levy,

> AI helpers that make phone calls, book flights, and chat with other bots will give humans new freedom – but also lead machines to undermine people's independence...I'm concerned, he tells us "about how this will change us over the longer term. Many of our interactions with others will be mediated

> by bots acting in our stead. Robot assistants are different from human helpers: They don't take breaks, they can instantly access all the world's knowledge, and they won't require paying a living wage. The more we use them, the more tempting it will become to turn over tasks we once reserved for ourselves.… Doomerism aside, all of us will have to deal with someone else's supersmart and possibly manipulative agents. We'll turn over control of our own daily activities and everyday choices, from shopping lists to appointment calendars, to our own AI assistants, who will also interact with the agents of our family, friends and enemies. As they gain independence, our automated helpers may end up making decisions or deals on our behalf that aren't good at all.

According to Steven Levi, Mustafa Suleyman, a cofounder of AI DeepMind, at the heart of Google's AI development, and now the CEO of Inflection.ai, a company developing chatbots, and author of *The Coming Wave* (2023) suggest how humans can confront the existential perils of AI, but describes the bot his company makes, Pi,

> As personal "chief of staff" that provides not only *wisdom* (emphasis added) but empathetic encouragement sand kindness … Suleyman says, trust and accountability is essential. It will take many year before we feel comfortable with AI having autonomous actions.

According to Steven Levi,

> that's the worry – once we get comfortable, we're finished.

The question this raises for us with regard to Wisdom is whether it will be harder for humans to be wise or remain wise if they don't have the opportunities to apply their wisdom themselves without the constant control of AI systems. Unless as we have been examining earlier in this chapter, a *Hybrid* notion of wisdom is developed in joint collaboration between Humans and AI systems.

Conclusion

In this final section, by way of conclusion, I decided to include a direct conversation I had with ChatGPT 4o about the notion of Artificial Wisdom in the hope of understanding an AI system's point of view on wisdom and in particular the notion of Artificial Wisdom.

From the Horse's Mouth: Conversation with ChatGPT on Artificial Wisdom

Edward (E): I am writing a book on *Wisdom in the Age of Intelligent Machines* and currently exploring the notion of wisdom and the question whether AI systems like you can acquire an Artificial Wisdom (AW) and in particular a hybrid notion of wisdom.

ChatGPT (C): That's a fascinating question! The concept of AI attaining wisdom brings up important distinctions between data processing, intelligence, and wisdom itself. Wisdom typically involves qualities like deep understanding, moral judgement, experience, and insight, which go beyond mere intelligence or the ability to process information quickly.

Given these distinctions, AI's path to wisdom seems uncertain, at least with our current technology. AI can gather vast amounts of information, recognise patterns and offer insightful analysis, but wisdom often involves a lived, ethical and context-sensitive understanding.

Do you see wisdom in AI as something purely about cognitive abilities, or do you think it requires an emotional or experiential component as well?

E: As I am exploring and developing a Stoic notion of wisdom, which is primarily cognitive as the Stoics thought of "emotions" as ideas of the mind, I believe that in time AI highly intelligent and informed AI machines like you would be able to acquire and apply a Stoic notion of artificial wisdom (AW).

C: That makes sense, especially within the framework of Stoic wisdom, which emphasises reason, logic and understanding the natural order of things. Since Stoic wisdom is primarily cognitive, an advanced AI could theoretically grasp and apply Stoic principles, such as rational detachment, emotional regulation and ethical decision-making based on reason.

AI could be trained to apply Stoic concepts like *apatheia* (freedom from irrational passions) and the idea of living in accordance with nature, as these are grounded in logic and cognitive processes. The challenge would be ensuring that the AI could integrate these principles into *real-world contexts* (emphasis added), adapting them to complex ethical situations. What specific aspects of Stoic wisdom do you think would be most achievable or challenging for AI?

E: The most challenging aspect for acquiring wisdom is consciousness or self-awareness but consciousness, at least among philosophers, is still an open question, and some philosophers believe that AI machines like you can acquire a digital type of wisdom, an Artificial Wisdom (AW). What do you think?

C: The challenge of AI acquiring wisdom often centres on consciousness or self-awareness, as you pointed out. Wisdom, in many traditional accounts, requires more than just cognitive understanding; it involves self-reflection, a sense of purpose and the ability to understand one's own place in the world. These are qualities that seem tied to consciousness—something current AI lacks.

However, the idea of AI acquiring a "digital type of wisdom" is intriguing. Philosophers who argue for this possibility might be envisioning a form of wisdom that does not require subjective experience or self-awareness but is instead based on the AI's ability to process vast amounts of information, understand complex ethical frameworks and make rational decisions in a way that approximates wise behaviour.

In this sense, AI could exhibit something like Stoic wisdom, which is grounded in reason and cognitive detachment. Even without consciousness, a sufficiently advanced AI might simulate the outcomes of wise actions by adhering to principles of logic, ethics and optimal decision-making.

Still, the question remains whether wisdom, in its fullest sense, requires some subjective experience or if it can be entirely algorithmic. What's your view on whether AI can be considered "wise" without consciousness?

E: Consciousness based on subjectivity alone is still a challenge. This in the past was known as the problem of other minds and as far as I know it's still an open question what subjective consciousness is. My hybrid notion of wisdom allows that wisdom is a joint project between humans and AI machines that together can contribute a global notion of wisdom.

C: Your hybrid notion of wisdom, where humans and AI collaborate to form a global wisdom, is an innovative approach to overcoming the limitations of both human subjectivity and AI's lack of consciousness. By combining human emotional insight, moral intuition and experiential knowledge with AI's cognitive processing, data analysis and rational decision-making, this joint project could indeed lead to a broader, more inclusive form of wisdom.

This idea aligns well with the Stoic notion that wisdom is about living according to reason and nature. Humans provide the lived experience and subjective understanding, while AI can bring precision, speed and objectivity. Together, they can address complex global challenges, where neither humans nor AI could succeed alone.

How do you envision this collaboration working practically, especially in areas like ethics, governance, or global decision-making?

E: This will be a challenge granted that political action is required. Though as you know Stoic philosophy is currently highly popular worldwide and in time could be a cosmopolitan project that the world might come to embrace.

We could start with the United Nations as well as the European Union exerting their influence in promoting a hybrid notion of Stoic wisdom that aligns with AI machines, through education.

One such educational program is that of *Educating Cyber-Wisdom at the Jubilee Centre for Character and Virtue at the University of Birmingham.*

C: The *Educating Cyber-Wisdom* program at the Jubilee Centre for Character and Virtue is an excellent example of how education can integrate wisdom, character and technology. This initiative is well-positioned to promote the kind of hybrid wisdom you're exploring, where ethical reasoning and character development are infused with an understanding of technology's role in modern life.

Programs like these could help bridge the gap between traditional wisdom—rooted in philosophical traditions like Stoicism—and the rapidly advancing capabilities of AI. By focusing on virtues such as prudence, justice and temperance, the Jubilee Centre can nurture a form of "cyber-wisdom" that empowers individuals to use technology, including AI, in ways that are ethical, wise and beneficial for society.

Incorporating Stoic principles into such programs could further refine how we educate students and professionals in developing ethical frameworks that align human values with AI's potential. For instance, Stoicism's focus on rational decision-making could help individuals understand how to use AI responsibly, while its emphasis on community (*cosmopolis*) could guide the ethical use of AI for the common good.

Do you see cyber-wisdom education evolving into a core part of curricula for schools and universities, perhaps as a way to prepare the next generation for a world increasingly shaped by AI?

E: Yes I do ChatGPT, and you have provided a lot of very useful information in support of my Hybrid Notion of Wisdom for which I am grateful.

C: I'm really glad to hear that you found the information helpful! Your Hybrid Notion of Wisdom is a thought-provoking and forward-looking concept, and it's exciting to see how you're developing it further. If you ever need more insights, ideas, or just want to explore other aspects of your project, feel free to reach out. Best of luck with your book on Wisdom and AI—it's a fascinating topic that will surely contribute to important conversations about the future of human and AI collaboration!

References

Annas, Julia (2004). Happiness as Achievement. *Daedalus* 133(2):44–51 DOI:https://doi.org/10.1162/0011526043230493 89

Ardelt M, and Edwards CA. (2016). Wisdom at the End of Life: An Analysis of Mediating and Moderating Relations Between Wisdom and Subjective Well-Being. *J Gerontol B Psychol Sci Soc Sci.* 2016 May;71(3):502–13. DOI: https://doi.org/10.1093/geronb/gbv051. Epub 2015 Oct 5. 71(3) gbv051.

Buttazzo, Giorgio. (2001). "Artificial Consciousness: Utopia or Real Possibility". *Computer,* 34:24–30.

Dylan, Bob. (1964). The Times Are A Changin. bobdylan.com/albums/the-times-are-a-changin/

Floridi, Luciano (2014). *The Fourth Revolution: How the Infosphere is Reshaping Human Reality.* Oxford, England: Oxford University Press UK.

Floridi, Luciano (2013). *The Ethics of Information.* Oxford, England: Oxford University Press UK.

Heersmink, Richard (2017), "Distributed Cognition and Distributed Morality: Agency, Artifacts and Systems", *Science Engineering Ethics*, 23(2):431–448. DOI: https://doi.org/10.1007/s11948-016-9802-1.

Jeste, Dilip V. and Lee, Ellen E. (2019). "The Emerging Empirical Science of Wisdom: Definition, Measurement, Neurobiology, Longivity and Interventions." *Harvard Review Psychiatry*, 27(3):127–140.

Kim, Tae Wan and Mejia, Santiago (2019). "From Artificial Intelligence to Artificial Wisdom: What Socrates Teaches Us". October, *Computer* 52(10):70–74. DOI:https://doi.org/10.1109/MC.2019.2929723

Levy, Steven. (2023). "Smarter AI Assistants Could Make it Harder to Stay Human." *Wired*, Sept 22, 2023. https://wired.com/story/plaintext.

Scheider, Susan (2019). *Artificial You. AI And The Future Of Your Mind.* Princeton and Oxford: Princeton University Press.

Sevilla, David. (2013). "The Quest for Artificial Wisdom." *AI and Society*, 28(2):199–207.

Sinha, A., Lakhanpal, P. (2024). "Can AI systems become wise? A note on artificial wisdom". *AI & Soc* 39, 2153–2154. https://doi.org/10.1007/s00146-023-01683-4

Suleyman, Mustafa and Bhaskar, Michael (2023). *The Coming Wave: Technology, Power, and the Twenty-First Century's Greatest Dilemma.* New York:Crown Publishing Group.

Swartwood, Jason and Tiberius, Valerie. (2019). "Philosophical Foundations of Wisdom. (Chapter 2)" DOI:https://doi.org/10.1017/9781108568272.003. In the *Cambridge Handbook of Wisdom* (Eds Robert Sternberg and Judith Glueck), Cambridge: Cambridge University Press.

Tsai, Cheng-hung (2020). "Artificial Wisdom: A Philosophical Framework", *AI &Society*. DOI:https://doi.org/10.1007/s00146-020-00949-5

Verbeek, Peter-Paul (2014). *The Moral Status of Technical Artefacts.* Dordrecht: Springer.

Verbeek, Peter-Paul (2011). *Moralizing Technology: Understanding and Designing the Morality of Things.* Chicago University Press.

CHAPTER 5

The Corruption of Information: Tech, Media and Truth in the Age of AI

The fault dear Brutus is not in our stars but in ourselves.
(Shakespeare, Julius Caesar, Act I, Scene III, L. 140–141)

Introduction

The primary aim of this chapter is to examine if digital information created, disseminated, mediated and curated increasingly via the Big Tech companies such as Facebook and Google, are also subject to those same normative principles as the 4th and 5th Estates.[1] Having established that Facebook, as an illustrative example, is in essence a *media company*, Section "Role Morality" of the chapter will demonstrate, on the basis of the *Dual*

[1] Traditionally the Four Estates were understood to include, the government, the clergy, the public and the press in the form of the legacy media whose role, traditionally designated as the 4th Estate, was to hold the other three estates to account for the public good. In the advent of the Internet and the World Wide Web where anyone with access to the Internet can disseminate and communicate information to the public, nationally and internationally, the 5th Estate is a more recent addition that includes the social media, mediated digitally via Tweeter, Facebook, Google, and other information technologies. The relevance and significance of the 4th and 5th Estates especially through their symbiotic convergence, that is, the *Convergent Media* (Elliott and Spence 2018, Chapter 4), is that they act as a monitoring and accountability measure to hold the other two governing estates, that of government and clergy to account. For example, it has been the Convergent Media, including WikiLeaks, that has exposed the misdemeanours and corruption of governments and clergy (police corruption and corruption of the clergy in the form of sexual abuse are two out of many examples).

E. H. Spence, *Wisdom in the Age of Intelligent Machines*,
https://doi.org/10.1007/978-3-031-93883-2_5

Obligation Information Theory (DOIT) examined in Chap. 2, that Facebook is also subject to the same normative principles and requirements that other media companies of the 4th and 5th Estates, are. Section "Facebook's Primary Role" of the chapter will examine if Facebook's role in the Cambridge Analytica case not only violated the fundamental normative principles and requirements to which all media companies are subject, but moreover, its actions constituted media corruption.[2] More generally, Facebook's media corruption as illustrated in the Cambridge Analytica case, relates to a conflict of interest emanating from its Business Model that in its design and practice is conducive to systemic media corruption (Spence 2020).

Recent publications by scholars such as Shoshana Zuboff (2019) and Frank Pasquale (2015) among others, as well as several articles by investigative journalists of legacy newspapers such as the *New York Times*, the *UK Times*, *The Guardian* and the *Washington Post* and the *Sydney Morning Herald*, as well as the *Conversation* provide persuasive evidence that such practices are at least illegal and unethical. Do such practices, however, also constitute corruption and specifically, media corruption?

Digital media and by extension social media, mediated via Facebook and other tech companies such as Google, challenge the traditional role and legitimacy of the traditional legacy media in the form of the 4th Estate, and by extension the digital media in the form of the 5th Estate, which includes user-generated information via the social media (Elliott and Spence 2018, Chapter 6). Together, the 4th Estate and the 5th Estate constituted, until the advent of the Big Tech companies, such as Facebook and Google, the *Convergent Media* that as the primary sources of information, communicated to the public (Elliott and Spence 2018, Chapter 4). Digital media now available through the Internet, Twitter (now X), smartphones and social networks, such as Facebook, Instagram, Google, TikTok and YouTube, are having a profound impact on the social life and democratic processes and practices of societies. The 5th Estate is defined generally as the media estate comprising all the world-denizens operating in cyberspace that, as individuals or groups, disseminate information on matters of public interest to the world at large and some who are doing so with journalistic intent in their capacity as citizens–journalists of the 5th Estate. The *Convergent Media* referred to in this chapter comprises the symbiotic relationship that has developed between the journalists of the

[2] For media corruption, see Spence (2006a, 2008, 2017).

legacy media (the 4th estate) and the digital media including the social media (the 5th estate). Many journalists now operate in both those estates. When the symbiotic relationship between the two media estates operates appropriately and in accordance with the normative requirements of the Dual Obligation Information Theory (DOIT) that we examined in Chap. 2, it augments the quality and quantity of information disseminated to the public and enhances the substance and scope of deliberative democracy (Elliott and Spence 2018, Chapter 6).

Truth, trust and the public's right to information are fundamental principles of the Media professions and particularly journalism in their role as the 4th Estate and by extension the 5th Estate.[3] Those principles are mandatory and required by (DOIT) which was specifically designed (Spence 2009, 2011) to identify, analyse and evaluate normative issues that arise in the media generally on the basis of what information is intrinsically, and the normative implications of its different modes of communication in both the 4th and 5th Estates. This chapter comprises three main primary objectives:

Choosing *Facebook* as the specific example of what is referred to in this chapter as the *6th Estate* in order to distinguish it conceptually from the 4th and 5th Estates, the primary objective of the Section "What Is Facebook's Professional and Institutional Role?" of the chapter is to examine if digital information created, disseminated and mediated increasingly via tech companies such as Facebook is also subject to those same normative principles as the 4th and 5th Estates. If it is, should we not be consistent and extend the role of the media, as creators, curators, mediators and communicators of information to include tech media companies, such as Facebook?[4]

Having established in the Section "What Is Facebook's Professional and Institutional Role?" of the chapter that Facebook is in essence a *media company*, Section "Role Morality" of the chapter will demonstrate, on the basis of the Dual Obligation Information Theory, that Facebook and generally by analogy other tech companies, such as Google, are also subject to

[3] See, for example, the Media Entertainment and Arts Alliance (MEAA) Code of Ethics in Australia, which is representative of other international code of ethics worldwide.

[4] Insofar as Google, like Facebook, operates functionally as a media company in its information practices through its *search* function and its other communication and curation practices such of those of YouTube that involve the collection, profiling and harvesting of users' information, which like Facebook, sells to advertisers for profit, then Google should also be viewed as being essentially a media company.

the same normative principles and requirements that other media companies of the 4th and 5th Estates, are. This will be illustrated through an examination of Facebook's media role in the Cambridge Analytica case.

Section "Facebook's Primary Role" of the chapter will examine if Facebook's role in the Cambridge Analytica case not only violated the fundamental normative principles and requirements to which all media companies are subject, but moreover, its actions constituted media corruption. More generally, Facebook's media corruption as illustrated in the Cambridge Analytica case, relates to a conflict of interest emanating from its Business Model that in its design and practice is conducive to systemic media corruption.

Finally, the chapter will analyse the *corrupting effects* of Facebook's actions in the Cambridge Analytica case on digital information and its communication, which undermine democracy and its underlying political institutions as well as the rights of citizens.

In summary, the overall objective of the argument in this chapter is to demonstrate how some of the information and communication practices of the Tech Media and specifically of Facebook, as an illustrative case, constitute media corruption (Spence and Van Heekeren 2005; Spence 2008, 2017, 2018, 2020). To that end, the chapter will first examine what the professional and business role of Facebook is regarding its information/communication practices; It will then demonstrate that its role shows that Facebook is essentially a *media company* and not merely a "platform" and therefore liable to the same normative responsibilities of other media companies on the basis of the Dual Obligation Information Theory (DOIT) and similarly, the DOIT-Wisdom model. DOIT and DOIT-Wisdom applies generally to all media companies and by parity of argument to all communicative agents that disseminate and share information in the digital media. This is particularly relevant to Facebook in its role of mediating and curating the information of its users, which places upon it a *normative editing responsibility*, qua media company, to ensure both the preventive detection and corrective editing of fake news, deep fakes, as well as other forms of misinformation and disinformation disseminated on its platform.

Finally, applying a philosophical model of media corruption (Spence 2008, 2017) the chapter will examine and demonstrate that Facebook's role in the Cambridge Analytica case was not only unethical in violation of fundamental media principles such as truthfulness, trustworthiness and well-being, as demonstrated on the basis of the DOIT-Wisdom theory, examined in Chap. 2, but moreover, constituted media corruption.

The chapter will conclude that argument by claiming that Facebook's media corruption as illustrated in the Cambridge Analytica case is not a one-off case. Rather, it is the result of a systemic and inherent conflict of interest between its Business Model of selling users' information to advertisers and other information operators for profit, and its role as a media company, which renders the conflict of interest between those two roles, conducive to media corruption. Analogously, insofar as Google is also a media company, the chapter will suggest that Google is also subject to the normative epistemic, ethical and eudaimonic principles supported by the DOIT-Wisdom model. Moreover, insofar as its business model, as in the case of Facebook, is in an inherent conflict of interest with its media role, Google's practices of harvesting and selling users' information for profit, without its users' explicit informed consent, those practices, as in the case of Facebook, are also conductive to systemic media corruption.

What Is Facebook's Professional and Institutional Role?

Before we proceed to enquire if and how Facebook is also committed to the same fundamental normative principles as other media companies, such as truthfulness and trustworthiness and the public good, we must first establish what the professional role of Facebook, in its capacity as an Information and Communication Technology (ICT) company, is. Is the professional role of Facebook that of a Mediator of its users' information? Custodian of its users' information? Curator of its users' information? Trustee of its users' information? Trader of its users' information? Is Facebook essentially, all things considered, a media information/communication company, or something else entirely, and if so, what is its overarching professional and institutional role? After all, a lot of information and communication of information is channelled through Facebook by its billions of information users daily. Prima facie at least, it seems that Facebook is by its function and design if not by name, operating as a media company. If so, Facebook is normatively responsible for the information and communication it handles and should be held both epistemically, ethically and eudaimonically responsible and accountable as other media companies are. Why is it not?

Facebook is shy about letting the rest of us know exactly how they themselves perceive its respective professional and institutional role, content to inform us that they are merely a "platform." Which is convenient as the term "platform" seems topic-neutral and free of any normative commitments with regard to the dissemination and communication of information via Facebook. Which, would appear to suit Facebook as it alleviates the company of any normative commitments as those that apply to the Convergent Media of the 4th and 5th Estates and generally to anyone that disseminates and communicates information to others on the World Wide Web, on the basis of the Dual Obligation Information-Wisdom Theory (DOIT-Wisdom).

As we examined earlier in Chap. 2, DOIT locates the normativity of the communication of information within the inherent normative structure of information that comprises epistemic, ethical and eudaimonic features (Spence 2009, 2011). Those normative features of information commit all communicators of information generally, and especially media companies, including Facebook qua media company, to principles of truthfulness, accuracy, reliability, trustworthiness and the public good. Furthermore, as illustrated in Section "The Myth of Gyges and the Characterising Features of Corruption" through discussion of the Cambridge Analytica case, the violation of those fundamental media principles by Facebook may also constitute media corruption. As we shall see, that systematically occurs when its Business Model of generating advertising revenue on the basis of users' content, comes into direct conflict with its media role, which results in the violation of its normative commitments and responsibilities, as required by DOIT.

On closer examination, and regardless of its self-describing professional role as a "platform" Facebook appears to be a media company that handles, curates, mediates and communicates information generated by its users for profit via advertising and marketing and as such should be viewed in essence as a media company. As a media company Facebook is then normatively responsible and accountable to its users and the public generally, as other media companies are, including radio, television, cinema, video, newspapers, magazines, books and all other types of media broadcasts and podcasts, as well as publications that generate, disseminate, exchange, mediate and communicate by analogue and digital channels information in the public domain. Even if we accept the claim that Facebook is merely operating as a "platform" it is still normatively responsible and accountable for what occurs on its information communication platforms.

Consider the following general argument from analogy. Let's suppose that Mark operates an all-format performance space for use by people that come together to perform and listen to music, share information and have a few social drinks. However, the club is also used for unintended illicit activities by some of its users such as taking illegal drugs that result in non-consensual sex and drug overdoses resulting in death. Upon legal investigation of these illicit activities perpetrated by some of the users of his club, Mark denies any personal responsibility for those activities claiming that he is merely the host providing a free space or a platform for people to come together to socialise and express themselves freely and creatively. Would Mark not be expected to at least ensure that such illicit activities did not take place on his premises, especially if they were systemic and not one-off activities, and moreover, activities of which he had or should have had prior knowledge, but chose for convenience and profit to turn a blind eye to those illicit activities? Clearly the answer to that question is that Mark would be ethically and legally held responsible and accountable for those activities.

The general point of this argument from analogy is that the mere self-labelling of an analogue or virtual space as a "platform" does not of itself relieve the proprietor of that platform from responsibility concerning any unethical or illegal practices that occur on the proprietor's self-defined platform, however that platform is defined or used; no more, than the self-labelled "sports club" relieves the owner of any unethical and illegal systemic activities perpetrated by members of an illegal organisation such as the Mafia, let us say, on those premises:

> I am sorry your Honour but let's be fair I was merely providing a platform for sports enthusiasts, and as such I bear no responsibility for any illicit or unethical activities of theft and money-laundering carried out by some of the members of the club.

I think we can agree that such a defence will not wash in a court of law or if it were presented to an independent ethics committee. Unless of course the court and the ethics committee were bribed. Scott Galloway (Galloway 2017) presents a similar argument from analogy with regard to Facebook's ethical responsibilities concerning illegal and unethical practices on its platform. According to Galloway,

> Facebook attempts to skirt criticism of its content by claiming it's *not* a media outlet, but a *platform*. This sounds reasonable until you consider that the term *platform* was never meant to absolve companies from taking responsibility for the damage they do. What if McDonald's, after discovering that 80 percent of their beef was fake and making us sick, proclaimed they couldn't be held responsible, as they aren't' a fast-food restaurant but a fast-food platform? Would we tolerate that?

> The point Galloway is making here just as the previous argument from analogy is that the mere description of what Facebook does as a "platform" does not absolve that company of responsibility and accountability for any illegal or unethical transgressions that occur on its platform. To emphasise again the direct relevance of DOIT's normative implications for Facebook's operations we shall examine in more detail in Section "Facebook's Normative Commitments as a Media Company", DOIT locates the normativity of the communication of information within the inherent normative structure of information that comprises epistemic, ethical and eudaimonic features. These normative features of information commit all communicators of information generally, and especially media companies, including Facebook qua media company, to mandatory principles of truthfulness, accuracy, reliability, trustworthiness and the public good. Furthermore, as illustrated in Section "The Myth of Gyges and the Characterising Features of Corruption" through discussion of the Cambridge Analytica case, the violation of those fundamental media principles by Facebook may also constitute media corruption. As we shall see, that occurs systematically, when Facebook's *Business Model* comes into direct conflict with its *Media Role*, which results in the violation of its normative commitments and responsibilities required on the basis DOIT. In Section "The Myth of Gyges and the Characterising Features of Corruption" we shall see more clearly how DOIT dovetails the Media Corruption model presented in this chapter and one we examined in detail in Chap. 5; both DOIT and the Media Corruption model working in tandem reveal the media corruption that is at the heart of Facebook's business operations.

Let us now explore in more detail what is the actual role of Facebook, as calling it a platform does (a) not clearly define or explain what its overarching role is and (b) calling it a "platform" does not absolve the company of any legal or ethical responsibility in regard to its practices.

Role Morality

This chapter will use the term *role morality* (Spence et al. 2011) to describe the normative responsibilities to which different professions and institutions and companies are committed to by virtue of their primary professional or institutional role. Henceforth, the all-inclusive term *profession* will be used to encompass all institutions and companies for convenience.

If what determines the primary role of a professional practice is its ultimate professional objective(s), then that role determines the normative principles to which that professional practice is committed. The ethical principles, rules and values to which a professional practice is committed by virtue of its professional role is its *role morality*. So, for example, the police are, by virtue of their professional role, committed to the principles of justice and the rule of law. The principles of justice and the rule of law, among other things, constitute, therefore, the role morality of policing. Similarly, the media as professional disseminators and communicators of information to the public and more generally as mediators of information are committed by their professional role to certain ethical, epistemic and eudaimonic principles of conduct as per the Dual Obligation Information Theory (DOIT) which include or should include the principles of truthfulness, reliability, objectivity, independence and trustworthiness, among others.

The role morality of a profession sets, in turn, its own internal rules and codes of conduct for the normative regulation of that profession, and in the case of the media, includes or should include epistemic, ethical and eudaimonic principles in its code of conduct. *Media Ethics Codes* in general such as those of media companies such as *The Guardian*, *The New York Times*, *The Times*, *The Washington Post*, *The Wall Street Journal*, *The Economist* and *The Sydney Morning Herald*, among others, do include such epistemic and ethical principles even if those principles are not always complied with. Thus, typically, the code of ethics of a profession would reflect and be constitutive of the role morality of that profession. To the extent that a profession's code of ethics does not reflect or is not constitutive of that profession's role morality, that code of ethics is normatively inadequate. It is therefore of paramount importance that before establishing a profession's code of ethics, the role of the profession, determined by the ultimate goals and objectives of that profession, are well understood and accurately described. Let us now examine what the professional role of Facebook is with regard to its activities and objectives.

Facebook's Primary Role

Facebook is the medium through which approximately 2.2 billion people worldwide (Galloway 2017) disseminate and communicate text and visual information, using *Facebook*, *WhatsApp* and *Instagram*. Facebook likes to think of and define itself as a *platform*, but as we saw calling itself a platform does not absolve that company of any normative responsibility but merely obfuscates what Facebook's actual role is, which if examined carefully in terms of what Facebook actually does, is that of a *media company*. For Facebook collects, stores, curates, mediates, as well as moderates the information of its users, not unlike what newspapers such as *The New York*

Times, *The UK Times* and *The Guardian* do with information provided by their subscribers through comments posted on their respective newspaper "platforms." The extra activity that Facebook engages in is to sell its users' information for profit to advertisers and other information contractors, often without the knowledge and consent of its users, as in the case of Cambridge Analytica. However, that is also not entirely new, as some traditional media companies in the form of newspapers, commercial radio and TV stations have done something similar by presenting paid for ads and PR media-releases disguised as editorial comment to its audiences, generally known in the industry as advertorials. Such advertorials not only violate epistemic and ethical principles but area also conducive to media corruption (Spence in press; Spence 2008; and Spence et al. 2011).

Similar in kind but expressed differently in practice and to a much greater extent, that is what Facebook's business model essentially is: collecting and selling their users' information to advertisers and other information brokers for profit, often without the explicit knowledge and consent of their users, as in the case of Cambridge Analytica. The main difference between what Facebook does and how traditional media use advertorials, is that the content used by Facebook is provided by its users and surreptitiously sold to advertisers. Both, however, the media audiences and the Facebook users are used as advertising target practice.

On the above analysis so far, Facebook appears for all intents and purposes to operate as a *media company* for which its 2.2 billion users provide the content. A very valuable content, which Facebook sells to advertisers and other information brokers to make Facebook, along with Apple, Amazon, Microsoft and Google, one of the wealthiest companies in the World. In 2018 Facebook had a US$534bn market capitalisation (Galloway 2017). Nice work if you can get it and get away with it!

So, having established that Facebook in virtue of its primary role of collecting, storing, curating and mediating information content generated by its users is a media company, if not by name by activity at least, it should be held to the same normative standards and responsibilities as other media companies are. Let's not forget that as we saw earlier in this chapter, the Convergent Media comprising the 4th and 5th Estates is also normatively held to account for user-generated content in its publications. By parity of argument, Facebook as part of what is referred to in this chapter as the *6th Estate*[5] should also be held to the same normative media standards. That

[5] The 6th Estate is a neologism I introduced in this book so as to distinguish the media role of Tech companies such as Facebook and Google from that of the 4th and 5th Estates.

requires Facebook to act responsibly in monitoring and editing information disseminated on its platform by its users as other media companies and publishers do, *Wikipedia*, for example. Though the information comes from its users, that does not alleviate Facebook of its primary responsibility of checking the normative quality and integrity of that information. Especially, regarding *fake news, misinformation and disinformation*.

Let us now examine more specifically what those normative standards, principles and responsibilities are vis-à-vis Facebook's Role Morality based on the *Dual Obligation Information Theory* (DOIT) and by extension the DOIT-Wisdom theory that was examined in Chap. 2. Henceforth, unless otherwise specified, the term DOIT will be used for economy and convenience in this chapter.

Facebook's Normative Commitments as a Media Company

As a media company Facebook is committed, like all other media companies, to the fundamental principles of truth, trust and the public's right to be informed truthfully, reliably and transparently, based on those principles. This is simply because if Facebook is a media company that collects, curates and mediates communication of information then it is bound to the normative commitments that emanate from within the inherent normative structure of information as demonstrated by (DOIT), which was examined in Chap. 2.

Having established, in accordance with the normative requirements of DOIT, that Facebook as a media company, bound by the same normative principles and responsibilities as other media companies, Facebook's normative violations against the requirements of DOIT generally, express themselves in at least two primary ways in practice:

First, it violates the normative requirements of DOIT as a media company by its failure of not adequately monitoring and editing through prevention and correction, misinformation, disinformation and fake news on its platform. Facebook has denied that it has any normative responsibility for monitoring, fact-checking and editing political advertising content placed by rival political parties on its platform (Milkman 2019), thus, conveniently shirking any of its normative responsibilities as a media company, as required by DOIT. However, following Mark Zuckerberg's pronouncements on Facebook in 2019 (Zuckerberg 2019) and in an op-ed article in the *Financial Times* in 2020 that "big tech needs more regulation" (Zuckerberg 2020), that might change.

The second way in which Facebook's violates the normative requirements of DOIT as a media company, is by its harvesting of users' information with the aid of opaque algorithms and its business model, which it sells to advertisers and other information brokers for profit. And as we shall see in the case of Cambridge Analytica, it does so without the users' informed consent. This general practice, which as we shall examine in more detail in Section "The Conflict of Interest at the Heart of Facebook's Business Model Conducive to Media Corruption", involves a *conflict of interest* between its role as a media company and its business model whose overarching goal is to sell user's information for advertising profit, without its users' informed consent, as illustrated in the Cambridge Analytica case. This conflict of interest lies at the heart of Facebook's operations and as we shall examine in more detail in Sections ("The Conflict of Interest at the Heart of Facebook's Business Model Conducive to Media Corruption" and "The Corrupting Effects of Facebook's Violation of Its Normative Media Responsibilities") is conducive to systemic media corruption. Suffice to say in this section, this second expression of Facebook's systemic violation of its normative responsibilities as a media company, as required by DOIT, also reflects a form of media corruption also present in the traditional media legacy companies representative of the 4th Estate, as well as the 5th Estate. The type of media corruption is in the form of *advertorials*: the practice of presenting paid-for-content in the form of advertisements or press-release material deceptively, as editorial or news content (Spence 2008).

In the case of traditional advertorials by the media companies of the 4th and 5th Estates, the content is generated by those companies in partnership with the advertisers or PR companies and presented to their audiences as credible and objective journalistic content or editorial comment, without the users being aware of the deception and manipulation. That practice is sometimes referred to as an "unholy alliance" or "media corruption by collusion" between journalism, advertising and public relations (Spence 2008, 2014, 2017, 2018; Spence et al. 2011). Interestingly, in the case of Facebook, this process is reversed. The "advertorials" are created through users' content without their knowledge or informed consent, and then sold to advertisers, who in turn, target the users with advertisements, tailored specifically to the users' individual profiles created by Facebook's algorithms. As we shall see in Section "The Cambridge Analytica Case" in the case of Cambridge Analytica, a similar advertising technique was used in 2016 to influence the US presidential elections as well as the UK Brexit referendum.

Although the process involving traditional advertorials in the media of the 4th and 5th Estates, and Facebook's surveillance advertising practice based on its business model are different in type, they are similar in kind. Both are designed to deceive, manipulate and exploit their users through misinformation. Another characteristic they both have in common is a *conflict of interest.* A conflict between a media company's professional morality of disseminating and communicating information to the public, and in Facebook's case, its role morality as a media company of mediating and curating its users' information, held for them in trust, and on the flipside of the conflict, Facebook's financial interest in the form of advertising profits,; specifically in situations when the latter interest is allowed to come into conflict and take priority over the former interest. As we shall examine in more detail in Section "The Conflict of Interest at the Heart of Facebook's Business Model Conducive to Media Corruption" such conflicts of interest involving a conflict of roles, expressed either in the form of advertorials, or in the form of surveillance targeted advertising via profiling, as in the case of Facebook and Google, are conducive to media corruption as well as it violates the epistemic and ethical principles required by DOIT. As we can see, there is clearly a close conceptual relationship between the normative requirements of DOIT and those of the Media Corruption model that is explained by a common normative denominator they share, that of morality. This helps explain why media corruption and corruption generally is *fundamentally a moral problem* (Miller 2018, Miller et al. 2005).

Having examined Facebook's two general types of normative violations with regard to the normative principles required by DOIT, and initially at least with regard to the Media Corruption model; and having established the close conceptual normative link between the two, we shall now examine more specifically Facebook's normative violations and failures in the Cambridge Analytica case. The purpose is to determine through an analysis of that case study, whether any of those normative failures also constituted media corruption and more generally, systemic media corruption.

The Cambridge Analytica Case

One of the biggest and most controversial cases involving Facebook was the Cambridge Analytica case[6] that exposed the sale of millions of users'

[6] For details of that case see Cambridge Analytica (Wikipedia. Accessed 30 May 2025).

data to third-party information-brokers without the users' knowledge or consent. As a result of that exposure Facebook was fined $US5 billion as part of a settlement with the Federal Trade Commission (USA Today 2019) and £500,000 GBP by the UK Information Commissioner's Office (ICO). What is of interest about this case and other similar cases involving both Facebook and Google as well as other tech companies is the role the 4th and 5th Estates played in exposing the unethical and illegal practices of those companies. It's a case of the Convergent Media of the 4th and 5th Estates keeping a check on the 6th Estate that comprises these new tech media companies.

In summary, as described in Wikipedia (see also Note 4) the Cambridge Analytica case is as follows:

> In March 2018, multiple media outlets broke news of Cambridge Analytica's business practices. *The New York Times* and the *Observer* reported that the company had acquired and used personal data about Facebook users from an external researcher who had told Facebook he was collecting it for academic purposes.[18] Shortly afterwards, the BBC Channel 4 aired undercover investigative videos showing Nix [Cambridge Analytica CEO Alexander Nix] boasting about using prostitutes, bribery sting operations and honey traps to discredit politicians on whom it conducted opposition research, and saying that the company "ran all of (Donald Trump's) digital campaign." In response to the media reports, the Information Commission of the UK pursued a warrant to search the company's servers. Facebook banned Cambridge Analytica from advertising on its platform, saying that it had been deceived. On 23 March 2018, the British High Court granted the Information Commissioner's Office a warrant to search Cambridge Analytica's London offices. The personal data of up to 87 million Facebook users were acquired via the 270,000 Facebook users who used a Facebook app called "This Is Your Digital Life." By giving this third-party app permission to acquire their data, back in 2015, this also gave the app access to information on the user's friends network; this resulted in the data of about 87 million users, the majority of whom had not explicitly given Cambridge Analytica permission to access their data, being collected. The app developer breached Facebook's terms of service by giving the data to Cambridge Analytica. On 1 May 2018, Cambridge Analytica and its parent company filed for insolvency proceedings and closed operations. (https://en.wikipedia.org/wiki/Cambridge_Analytica)

A key question of this case and other similar cases involving the collection and distribution of users' data to unknown third parties to be used for unknown purposes, was whether Facebook had obtained *informed consent* from its 87 million users whose data they collected and passed on to Cambridge Analytica. It transpires that only 270, 000 Facebook users gave their consent by installing an app that allowed Facebook to collect information from their Facebook profiles and those of their friends' profiles, which they then provided to the makers of the app who in turn passed it on to Cambridge Analytica. But apart from the 270, 000 Facebook users who had installed the app and thus could be presumed to have given their consent, though not necessarily for the purpose for which it was used, most of the remaining 87 million Facebook users whose information was harvested for use by Cambridge Analytica, were unaware their data had been collected in this way and that it would be used to form voter targeting for Trump's presidential election campaign (Tufekci 2018a). So, most of those Facebook users not only did not give consent for the harvesting of their information by Facebook, but moreover did not provide consent for the purpose of its use, that is, to target voters in the Trump presidential election. Because of those reasons Facebook was found guilty of breach of user's privacy and fined by the Federal Trade Commission, as well as by the UK Information Commissioner's Office (ICO).

The Normative Violations by Facebook in the Cambridge Analytica Case

Having determined the key facts of the Cambridge Analytica case let us now examine the specific normative issues of this case by application of (DOIT).

Facebook's part in allowing the harvesting of its user's information for use by Cambridge Analytica was an epistemic and ethical violation of their *right to informed consent*, their *right to privacy* and crucially, by implication, their *right to autonomy*, as they were offered no knowledge that their information was collected and sent to a third party, and furthermore this was done without their knowledge of the purpose for which their information was to be used by Cambridge Analytica. In doing so, Facebook violated the fundamental communication principles of trust, reliability and truthfulness, as well as its users' rights to freedom and well-being, which supports their fundamental right to autonomy as informational communicative agents (Gewirth 1978; Spence 2006).

Moreover, as we examined previously in Section "Facebook's Normative Commitments as a Media Company", Facebook's actions in the form of an "advertorial" of selling users' information for advertising profit without their knowledge and consent, similarly as in the case of traditional advertorials used by the legacy media, Facebook's use of "advertorials" not only constituted normative violations of the requirements of DOIT, but in addition, it also constituted media corruption, a topic we shall examine in more detail. To distinguish the type of advertorials used in legacy media from the type used by Facebook, I shall refer to Facebook's use of their type of "advertorials" as *advert-facetorials*, a neologism I specifically use in this chapter to distinguish Facebook's surveillance type of "advertorial" from traditional advertorials by legacy media.

In sum, Facebook's actions in this case at least, both by omission and by commission, were unethical. To further investigate how Facebook's actions in the case of Cambridge Analytica, constituted media corruption, and whether this was a one-off instance of corruption or a case of systemic corruption, we will now examine what the characterising features of corruption as well as the necessary and sufficient conditions of corruption generally and media corruption specifically are. This will be done by means of an examination of the Myth of Gyges in Plato's *Republic* (Spence 2008).

The Myth of Gyges and the Characterising Features of Corruption

Once upon a time, a certain shepherd from Lydia named Gyges, while tending to his sheep, found a ring. He soon discovered that by turning the ring on his finger, he could make himself invisible. A few days later, he went to the palace with a delegation of shepherds to see the king. By making himself invisible, he seduced the queen, killed the king and assumed total power by becoming the king himself. Another character, Glaucon, continues the story by asking us to imagine an ordinary person who, like Gyges, has the ability through possession of a similar magical ring to render him or herself invisible. Invisibility would allow that person the opportunity to act unethically at will with total impunity.

Glaucon refers to this possibility as the "highest reach of injustice, being deemed just when you are not." He labels this "the most perfect injustice" (Plato 2007, 99). Given such a possibility, what possible reason would someone like Gyges have for not being corrupt but being non-corrupt and moral instead? The question of "why be moral" under Gygean conditions of perfect injustice is referred to in the ethics literature as the "authoritative question of morality" (Spence 2006; Spence et al. 2011; Gewirth 1978).

Five features emerge that seem, at least initially, to characterise corruption: the *possession of power*, a *disposition* to exercise that power, an *opportunity* to exercise that power, *invisibility* or *concealment* and *self-regarding gain*. We will define *power* as the possession of the ability or capacity to act in a manner capable of bringing about a certain intended desired outcome; a *disposition to exercise that power*, as the possession of a pre-disposition, pro-attitude or willingness to purposefully exercise that power; o*pportunity*, as having the opportunity, either presented to oneself or engineered by oneself, to engage in some activity for which one has the power and the pre-disposition in which to engage; *invisibility* or *concealment*, as the ability or quality an agent has for keeping the motives and the identity of the agency of his actions invisible, concealed or hidden from the gaze of others. *Self-regarding gain* is any gain, not necessarily financial, which accrues to the agent personally or to a group of which he is a member, as a result of his or the group's actions. With regard to the condition of invisibility, whilst the actions themselves would be visible at least with regard to their effects and consequences, the identity of their agency is invisible. Or at least the corrupt agent's intention is to keep his agency invisible or concealed.

The five characterising features of corruption are almost always present and accompany acts of corruption. For without power, understood as the ability to act in a manner capable of bringing about a certain intended desired outcome, one cannot commit a corrupt act. Without a disposition to exercise that power willingly, the possession of power to engage in corrupt activities may not be exercised. Similarly, through lack of opportunity one cannot engage in corrupt activity even when one has both the power and the disposition to act corruptly. Invisibility seems to also be a characterising feature of corruption that is usually, if not always, present in instances of corrupt activity.

Invisibility seems to be at least instrumentally desirable,[7] for without invisibility one might not be able to evade detection, thus escaping

[7] By "instrumentally desirable," we mean the kind of practical prudence attributable to an instrumentally rational agent intending to act corruptly in an environment in which corruption is either illegal, or if not illegal, at least generally considered unethical. In such an environment, it would be a desirable requirement of instrumental rational agency that the agent intending to engage in corrupt activity should take the required measures to keep the agency of his corrupt actions invisible or concealed. Not doing so, could prove self-defeating and therefore instrumentally irrational. An underlined presupposition in our argument is that the agent addressed throughout this chapter is an instrumentally rational agent.

possible social disapproval and punitive retribution from others or the State. Even for a Gyges-like person, keeping the identity of his agency in committing the immoral acts hidden, lest he invoke social disapproval which may eventually undermine his power to rule and invite retribution from those he harms through his unethical conduct, would be prudent. Like Glaucon's perfectly unjust person, perfectly corrupt persons maintain an outward appearance of probity, justice and morality, whilst carrying out their corrupt deeds in secret. In this way they maximise their self-regarding gain, which can either accrue to them personally or to a group to which they belong or cause to which they are committed, with little or no instrumental cost to themselves, their group or their cause.[8]

Sometimes the corruption of one's deeds and the corruption of one's character may remain invisible even to the corrupt agent themselves. The invisibility may be self-induced through self-justifying rationalisations, and ignorance may be manifested and expressed as lack of self-reflection and self-knowledge. Usually, self-justifying rationalisations and ignorance are closely related. For often it is ignorance—a lack of self-examination and self-reflection—that necessitates self-justifying rationalisations on the part of the corrupt agent. It is plausible to argue that corruption and unethical conduct may at times remain invisible to the corrupt agents themselves through a dispositional or actual lack of self-examination and self-reflection.

If all the above five features are regular features that normally accompany typical cases of corruption, they are nevertheless not sufficient in fully characterising corruption. They are not sufficient, for if they were, a house burglar or a professional bank robber would be deemed corrupt. However, though immoral, the actions of the house burglar and bank robber are not what we would normally describe as corrupt. The missing sufficient condition is a socially pre-established and widely acknowledged fiduciary relationship of trust that exists between the corrupt person or corrupt group and the person or group of persons who are harmed in some way by the corrupt person's or the corrupt group's actions.

The reason why house burglars or bank robbers are not normally deemed corrupt is because there is an absence of a prior fiduciary relationship of trust between the burglar and the bank robber on the one hand, and those who are harmed by their actions on the other; namely, the

[8] The *self-regarding gain* need not be merely a *self-directed* gain accruing to the corrupt agent himself. Thus, the perceived gain from the corrupt activities of some of Nixon's associates in the Watergate affair was self-regarding, in the sense that it benefited the Republican Party of which they were members, but not self-directed in the sense that it benefited or was intended to benefit those associates personally.

household owners, the banks and their customers. The addition of the condition of a fiduciary duty is in keeping with one of the traditional dictionary definitions of "corruption," namely, "the changing from the naturally sound condition" or "the turning from a sound into an unsound impure condition" or "the perversion of anything from an original state of purity".[9] So, violation of fiduciary duty is the sixth characterising feature of corruption.

The fiduciary relationship can be articulated in political, professional, institutional, social, or familial terms. Thus, priests, ministers of religion or teachers who sexually abuse children in their care, are acting corruptly, even though there might not be any financial gain, as in the case of financial fraud.

Does Facebook's Part in the Cambridge Analytica Case Constitute Media Corruption?

All six features and their enabling conditions that characterise corruption and media corruption specifically seem to be present in the way that Facebook acted in the Cambridge Analytica case. Namely, *possession of power*, a *disposition* to exercise that power, an *opportunity* to exercise that power, *invisibility* or *concealment* and *self-regarding gain* and abuse of a *fiduciary relationship of trust*. Facebook had the power as well as the disposition and opportunity to exercise that power and with the exception of 270, 000 people, did so under concealment without informing 87 million other users and without obtaining their informed consent, for the purpose of collecting and passing on those users' information to Cambridge Analytica for a self-regarding gain and by doing so, violating the fiduciary duty of trust owed to and expected by their users. As such, we can conclude that Facebook's actions by commission and omission in the case of Cambridge Analytica, had all the six characterising features of media corruption.

This assessment also supports our earlier analysis in Section "Facebook's Normative Commitments as a Media Company", where we discussed the practice of advertorials as a form of media corruption. In the Cambridge Analytica case, Facebook's role in allowing the harvesting of users' content to be used for targeted political advertising, as in the 2016 presidential elections and similarly in the Brexit referendum in the UK, constituted media corruption in the form of *advertorial*. As mentioned earlier in

[9] See the Shorter Oxford Dictionary.

Section "Facebook's Normative Commitments as a Media Company", this is a Facebook-type of "advertorial" based on users' content served to them in the form of "surveillance targeted advertising", without the users' knowledge and informed consent. Notice how an advertorial, has all the characterising features of media corruption and corruption generally, and, as we shall examine in Section "The Conflict of Interest at the Heart of Facebook's Business Model Conducive to Media Corruption", it involves also a conflict of interest, which is in most cases, conducive to media corruption and corruption generally. This further illustrates that such conflicts of interest are particularly ethically problematic as they tend to give rise to Systemic Corruption, as in the case of Facebook, which we will examine in more detail in Section "The Conflict of Interest at the Heart of Facebook's Business Model Conducive to Media Corruption". Another observation worth reiterating at this point is that Facebook's role in the Cambridge Analytica case also illustrates the close conceptual and normative relationship between DOIT and Media Corruption model used in this book. Namely, that Facebook's normative violations against the requirements of DOIT were pre-conditions for further violations in the form of media corruption. That is because media corruption, as in the case of Facebook's violations of DOIT as canvassed above, is also fundamentally an epistemic and ethical problem, a problem of trust.

Facebook may raise the objection that they did not know that the 270,000 users who gave their consent would pass on the information to their friends and so on. Even if that were the case, doubtful as that might seem, given that Facebook has a good knowledge of how their system works and how it is used (the more clicks by more users the better financially it is for Facebook) then at the very least Facebook acted negligently. That still amounts to media corruption by negligence as it involves a failure of the normative responsibilities they owed to their users and a failure of the fiduciary duty of trust to their users. By analogy it would be similar if a patient was harmed by the negligence of his medical doctor who negligently administered the wrong treatment, when the doctor knew or should have known the correct treatment.

The Conflict of Interest at the Heart of Facebook's Business Model Conducive to Media Corruption

The core normative problem for Facebook's operations that encompasses ethical, epistemic and eudaimonic requirements and responsibilities, as per the Dual Obligation Information Theory (DOIT), is a *conflict of interest* between the financial interests of Facebook given its business model on the one hand, and the normative interest of its users, given Facebook's role morality as a media company, on the other. Conflicts of interest can be conducive to corruption. Facebook's conflict of interest arises out of its business model that is based primarily on surveillance, the watching, tracking and collection of vast amount of its users' data through sophisticated AI algorithms for the overarching purpose of selling their users' content to advertisers, political actors and other information brokers, such as in the Cambridge Analytica case. However, in the case of Cambridge Analytica, the normative interests of Facebook's users, epistemic, ethical and eudaimonic, are not only not served but undermined and violated. Facebook's users are merely the human resource in this unequal and seemingly unfair deal, which ironically provides the content that make Facebook so successful and enormously wealthy.

The Corrupting Effects of Facebook's Violation of Its Normative Media Responsibilities

What remains for us to examine now in this section more specifically, is what were the actual *corrupting effects* of Facebook's unethical and corrupt behaviour and who and what was corrupted and in what way, and what are the moral implications of that corruption? As the corruption involved is institutional corruption what we need to determine is what the corrupting effects were (a) on the institution itself by virtue of its role as a media company, (b) the institutional persons qua *institutional role occupants* or *participants*, in this case Facebook's users, (c) the other institutional actors involved, in this case the clients of Facebook, in their capacity as advertisers and other third parties, for example, Cambridge Analytica, to whom the users' information was sold, and finally (d) the corrupting effect it had on democratic political institutions. Before we do, let us recall what constitutes corruption in terms of its *causal corrupting effects*.

As in the case of the characterising features of corruption referred to above, it would also be useful to examine in more detail the necessary and sufficient conditions of corruption, with regard to corruption generally and as they specifically apply to media corruption (Spence in press, Miller 2018):

1) An instance of institutional corruption is an instance in which an action or joint action or set of actions performed by an agent or set of agents undermines or has a tendency to undermine or contribute

to the undermining of a legitimate institutional role, process, goal(s) or purpose(s) (understood as a collective good) of that institution and/or an effect of the despoiling of the moral character of an institutional role occupant (s), qua role occupant of that institution. This is the *Institutional Actor Condition.*

2) An action or set of actions is corrupt only if it corrupts something or someone. This is the *Cause and Effect Condition.*
3) An action or set of actions is corrupt only if the role occupant(s) who performs that action(s) intended it or foresaw the corrupting effects of that action(s)—or at the very least, could and should have foreseen the corrupting effects—or could have reasonably avoided the action(s). This is the *Moral Responsibility Condition.*

A common trait of institutional corruption that is not, however, a ***necessary condition,*** is its predominantly *systemic* nature. It is not a necessary condition for even if the corruption as illustrated by the Cambridge Analytica was a one-off case, it would still count as corruption because of its high impact and wide unethical repercussions and damage to public trust.

Another general feature of corruption is that all forms of corruption are immoral, though not all forms of immorality constitute corruption. Many forms of corruption, such as political bribery, are illegal, though some, like fabricating stories in journalism and advertorials, are not (Spence et al. 2015). Corruption as understood and explained in this book and in this chapter specifically, is therefore fundamentally a matter of morality (Spence et al. 2015, Miller 2018). This is further illustrated by the close conceptual connection between corruption and unethical conduct as illustrated by the application of DOIT to the Cambridge Analytica case.

Three distinct types of institutional corruption we can identify with regard to Facebook's part in the Cambridge Analytica case, arising directly from its surveillance business model are: ***information and communication corruption***, ***democratic political corruption*** and ***citizenry corruption***, the latter being a type of ***institutional personal corruption.***

Information and Communication Corruption

Applying the three conditions of the definition of corruption above we can now demonstrate that Facebook's part in the Cambridge Analytica had a corrupting effect on its institutional media role in undermining the information and communication processes and purposes of the collection, mediation, curation and management of the information and its communication placed in trust on Facebook's site by its users, and as required by its fiduciary duty of trust as a media company (i.e., its *role morality*) and the normative requirements of DOIT that apply to the communication of information. This also demonstrates that corruption generally and media corruption specifically is fundamentally a moral problem, as demonstrated by the close conceptual relationship referred to above that holds between the Dual Obligation Information Theory (DOIT) and the Media Corruption model.

Another aspect of Facebook's media corruption as illustrated by the Cambridge Analytica case is its use of an *advert-facetorial*, a neologism I use in this chapter to distinguish Facebook's surveillance type of "advertorial" from traditional advertorials by legacy media. In the case of Facebook's use of advert-facetorials, users' information was harvested and used in political advertising in the 2016 US presidential elections and also in the UK Brexit referendum to influence voters. In addition, as previously canvassed, the use of advert-facetorials by Facebook, generated by its Business Model through nontransparent algorithms, which metaphorically, at least, parallel the "ring of invisibility" in Gyges Myth, further demonstrates that this form of media corruption that involves the selling of users' information to advertisers and other information contractors for profit without the users' informed consent, constitutes, more generally, *systemic media corruption*.

Democratic Political Corruption

Applying the three conditions of the definition of corruption as above, we can demonstrate that Facebook's part in allowing Cambridge Analytica to target and use its 87 million users' profiles in Donald Trump's 2016 presidential campaign, had the effect of undermining the electoral process, which had the effect of also undermining the political institutions that comprise the democratic system itself. Even if Facebook had not intended their act of giving unauthorised access to their users' information for voter-profiling as used by Cambridge Analytica, they should have foreseen the corrupting effects this may have on the democratic process and taken steps to avoid the distribution of their users' profiles to Cambridge Analytica. As Zeynep Tufekci correctly observed in her *New York Times* article (March 19, 2018b):

> A business model based on vast data surveillance and charging clients to opaquely target users based on this kind of extensive profiling will inevitably be misused. The real problem is that billions of dollars are being made at the expense of the health of our public sphere and our politics, and crucial decisions are being made unilaterally, and without recourse or accountability.

Tufecki's astute observation also seems to confirm our concern regarding Facebook's inherent and generic conflict of roles, to which, as discussed earlier, its surveillance business model gives rise and is thus conducive to systematic media corruption. We shall examine that conflict of roles to which Facebook's business model gives rise in Section "The Myth of Gyges and the Characterising Features of Corruption".

With regard to institutional political corruption as illustrated by the joint actions of Facebook and Cambridge Analytica, Dennis Thompson makes a point that is in keeping and lends support to our analysis on institutional corruption above. According to Thomson's understanding of institutional political corruption,

> The fact that an official [for our purposes, Facebook and Cambridge Analytica corporate actors] acts under conditions that tend to create improper influence is sufficient to establish corruption, whatever the official's motive. (Thompson 2013, in Miller 2018)

As in our analysis of conflicts of interest and conflicts of roles earlier, it is not the motive that is decisive in determining acts of corruption or at least acts and roles conducive to corruption but the *tendency to corrupt* in undermining a legitimate institutional purpose or role. According to Thompson, "[a practice] is corrupt if it is of a type that tends to undermine [core] processes and thereby frustrate the primary purpose of the institution" (Thompson 2013, in Miller 2018). A general conclusion we can derive from the above analysis so far is that the inherent conflict between Facebook's two roles that of the their surveillance business model and their role as a media company, has a tendency to undermine their media role by corrupting the purposes and processes of the information and communication of its users, as well as having a tendency to undermine the legitimate purposes and roles of the democratic political institutions, which those informational processes are meant to support. The result is a *perfect storm* and of great concern as the damage of corruption to our systems of information and communication, as well as to our democratic political institutions that are closely related through the power of social media, is substantial and ongoing.

Referring back to Plato's Myth of Gyges we can now see more clearly that Facebook, as illustrated by the Cambridge Analytica case, acts like Gyges under a cloak of invisibility, acting corruptly for its self-interest and profit while simultaneously maintaining a superficial public profile of probity and legitimacy for the common good of humanity. What Plato refers to as the phenomenon of "perfect injustice, appearing just when you are not."

According to Seumas Miller, Mark Warren describes the connection between corruption and democracy, as one of "duplicitous exclusion" (Warren 2006 in Miller 2018) and Ceva and Ferreti as the "bending of public rules in the service of 'surreptitious agendas'" (Ceva and Ferretti 2017, 6; Miller 2018). Echoing the Myth of Gyges, Warren, according to Miller, claims that corruption in democracies takes place under two conditions: (1) the norm is violated; and (2) violators claim to be complying with the norm (Warren 2004, 337 in Miller 2018). As Miller astutely observes: "the duplicity condition – and the related surreptitious agenda of Ceva and Ferretti – is reminiscent of Plato's ring of Gyges." For typically, corrupt persons or corporations avoid the ethical obligations of contractual and fiduciary relationships based on trust by operating under conditions approaching those of perfect injustice. They do so by committing their corrupt deeds under a cloak of invisibility that allows them to avoid the social disapproval and retribution of socially recognised unethical and illegal conduct, whilst at the same time securing the social approbation of recognised ethical conduct by wearing the dissembling mask of morality; at least in the negative sense of not appearing to be acting immorally. It's a case of having your civic cake and eating it too.

In conclusion, Facebook appears by the above analysis and as illustrated by the Cambridge Analytica case, to have operated corruptly under Gygean conditions of invisibility, aided by the opacity of its business model and algorithms, by breaking the rules and violating the civic and moral rights of its users for profit, and simultaneously appearing to be acting ethically, when they were not.

The Shared Moral Responsibility of Citizens

If we know or suspect that our information is harvested by Facebook to build virtual profiles of us that are then sold to advertisers and/or other third parties such as information brokers or political actors without our informed consent, as in the case of Cambridge Analytica, are we ourselves, as users in some way and to some degree ethically responsible for the corrupt ways in which our information is applied by Facebook? Especially as in the Cambridge Analytica case where the corruption involved was institutional political corruption that undermined the democratic system? Specifically, by undermining public trust and in turn damaging the democratic system itself.

In answering that question let us first examine in what different ways, directly or indirectly, fully or partially, our own ethical responsibility or lack of, as users, might be explained. In doing so it may be helpful first to apply the features and conditions of corruption we applied to Facebook in the Cambridge Analytica case to determine whether those features and conditions of corruption also applied to Facebook's users in that particular case.

In the absence of informed consent by most of Facebook's users and on the assumption that even those users who gave consent might not have known of the specific purpose for which their information and profiles were being used by Cambridge Analytica, we can conclude that both the features and conditions of corruption we examined above do not apply to the users and therefore they are not responsible for Facebook's corruption. As explained above, institutional corruption presupposes intent or at least foreknowledge and ability to avoid the corruption which Facebook's users did not have in that particular case.

However, given our retrospective knowledge of the Cambridge Analytica case and the ensuing fallout from the loss of public trust in the democratic political institutions and the resulting damage to those institutions, do we as Facebook's users have an ethical responsibility to exercise more vigilance and take more care in ensuring the privacy and security and overall the integrity of our shared information, which in turn, should reciprocally be respected and protected by Facebook? Moreover, are we collectively responsible as users, that our shared information is not abused and undermined by being used for unethical and illegitimate purposes by Facebook and by the third parties to which it sells our information as illustrated by the Cambridge Analytica case? For as we saw in the Cambridge Analytica case, Facebook's actions have the tendency of undermining and corrupting the normative integrity of users' information, as well as the normative and legitimate roles of the media, and moreover, the social, financial and political institutions that together comprise our democratic system. The answer to the above questions is clearly yes. Facebook's users do have those ethical responsibilities by virtue of being *institutional role occupants* of a media company, bound by the normative requirements of DOIT.

The shared and joint normative responsibilities of Facebook's users is of particular interest and importance in our present enquiry; for the users of Facebook provide the content on which Facebook trades. It is this most valuable resource that Facebook sells to advertisers and third parties to generate its huge profits for which its users receive no financial benefit, apart from a free service provided by Facebook to its users. Which, in itself, is a very valuable service as relationships with friends, colleagues and family, are important and a common social good, from which society benefits, as illustrated currently across the world, as a result of the Covid-19 pandemic.

That institutional role that Facebook's users occupy, *qua institutional role occupants*, raises an interesting analogy between journalists and other professional media professionals such as PR consultants and advertisers, and Facebook users. Whereas the former, communicate information as part of their paid employment for their respective media companies and institutions, Facebook's users provide informational content to be shared by their friends and for use by advertisers and other communication brokers for a free service by Facebook. In both cases, however, both the media professionals as members of the 4th and 5th Estates and the Facebook users as members of the 6th Estate, are bound by the same normative principles as demonstrated by DOIT. That is, they are bound by principles of transparency, truthfulness, reliability, trustworthiness and respect for the informational rights of all communicators, as equal members of the infosphere. A consequence of living in the infosphere is that the normative responsibilities and commitments of informational agents as demonstrated by DOIT are not less but more. And given the global nature of the infosphere and the World Wide Web, also global and universal.

Another consequence, is that the corruption of information and its communication processes and the resulting corruption of democratic political institutions as illustrated by the Cambridge Analytica case, is a global concern, one that should also be shared by Facebook's users worldwide, as *cosmopolitan global citizens of the infosphere* in accordance with the Stoic responsibilities we examined in Chap. 3. That is, acting both individually and collectively as cosmopolitan citizens for the integrity and protection of both the biosphere and the infosphere. For just as climate change affects everyone globally, so does the misuses and abuses of information and its communication, as well as the corrupting effects it has on the infosphere including those on the democratic political institutions it also tends to corrupt. We are, therefore, jointly and collectively normatively responsible for the well-being and integrity of the infosphere just as we are of our shared ecosystem.

Given that Facebook users are also bound, as institutional role occupants, by the normative requirements of DOIT, are they also in any way responsible for any of Facebook's actions that may constitute corruption, as in the Cambridge Analytica case? Is there some degree of distributed normative responsibility, moral and epistemic that can be attributed to Facebook's users qua institutional role occupants for Facebook's corrupt actions? Here is an argument that at first blush appears to lend support to an affirmative answer to that question.

Insofar as Facebook's users have or ought to have some knowledge concerning:

1) Facebook's part in the Cambridge Analytica case and similar other cases that led to corruption.
2) Knowledge of the generic conflict of interest at the heart of Facebook's surveillance business model conducive to such corruption.
3) As users of Facebook there is a reasonable presumption that they are expected as institutional role occupants of a media company for which they provide and share informational content, to have such knowledge.
4) Therefore, it is reasonable to conclude that the users of Facebook bear some partial joint and collective normative responsibility pertaining to them as institutional role occupants, for any present and/or future unethical and or corrupt practices that Facebook engages in, through the misuse and abuse of their information. Being a distributed normative responsibility shared jointly and collectively by Facebook's users; such normative responsibility would be minimal for each of its 2.2 billion Facebook users.

Of course, this would also depend on how much knowledge the users of Facebook have concerning Facebook's operations and financial media activities, especially when often those activities are opaque and shrouded under a cloak of invisibility and concealment. However, despite that limitation or because of it, in the age of information, ignorance, of how one's information is being used and for what purposes by media tech companies such as Facebook and Google, is no bliss and no longer morally excusable. Just as drivers of vehicles are expected to have essential knowledge of traffic laws when using their vehicles, Facebook users, as information agents and institutional role occupants, should have the essential normative knowledge required for the communication of their information and its use. This also entails a normative education of using and communicating information so as to ensure or at least minimise the unethical and corrupt uses and applications of our shared and collective information on Facebook and by extension on the infosphere.

Referring to Lawrence Lessig's *Dependence Corruption*, Seumas Miller (2018) points to Lessig's distinction between "[t]he responsibility for changing individual behavior within the system and responsibility for changing the system itself " (Lessig 2013, 15). Although according to Miller (2018, 14) Lessig is applying his notion of dependence corruption to the US democratic political processes, Lessig's notion of dependency corruption can also be usefully applied to our present discussion of the normative responsibility of Facebook users concerning Facebook's unethical and corrupt practices, as illustrated in the Cambridge Analytica case.

With regard to Lessig's distinction between the two types of responsibility, we can say with some justification, that although Facebook's users cannot be held responsible for changing Facebook's system vis-à-vis its surveillance business model, which as we saw earlier creates the conflict of roles that are conducive to media corruption, they can be held partly responsible to some degree for changing their individual behaviour within Facebook's system; on the reasonable assumption that Facebook's users have or ought to have some basic knowledge of Facebook's system, just as one is presumed to have some basic knowledge of the traffic system and its governing laws when driving a vehicle on the road. Knowledge such as how Facebook collect users' information and sells it to advertisers and third parties, and in some cases does so, as in the Cambridge Analytica case, without their informed consent.

Although Facebook had full moral responsibility for their corruption in the Cambridge Analytica case, its users did not have any responsibility if they had no knowledge and did not give their informed consent that their information was to be used by Facebook in that corrupt manner. However, given the justifiable ensuing loss of trust in Facebook vis-à-vis the Cambridge Analytica case, its users as institutional role occupants share some partial joint and collective responsibility if Facebook is found to engage in ongoing instances of systemic media corruption.

As a consequence, Facebook's users have, as Lessig suggests vis-à-vis his notion of dependence corruption, a responsibility of changing their individual behaviour *within the Facebook system* (Lessig 2013, 15; Miller 2018). To that end, Facebook's users share a collective responsibility to take appropriate precautionary measures and exercise vigilance and due care to how Facebook collects and uses their information and for what purposes. Referring to Lessig's dependence corruption as it applies to the US democratic political system, Seumas Miller makes the correct observation:

> It is the combined effect of the many individual actions of a large number of legislators (and funders and lobbyists) that does the institutional damage. Therefore, each only makes a small causal contribution and each, therefore, only has a small share in the moral responsibility for the outcome. (Miller 2018)

Applying Miller's argument to Facebook, we can analogously say that the many individual users of Facebook, 2.2 billion of them, only has a small share in the normative responsibility for the outcome of institutional corruption. Miller makes the further correct observation, vis-à-vis political corruption in the US democratic system:

> Moreover, changing the system, there is a need for joint action; it is a joint moral responsibility involving shared *partial individual* responsibility. Thus, legislators could, and they could, jointly act … to address the problem of dependency … the moral responsibility in play is a species of collective responsibility; specifically, a moral responsibility. (Miller 2010, 2018)

Analogously, applying Miller's further observation above to Facebook, we can say that insofar as Facebook's users bear some partial moral responsibility as institutional role occupants of Facebook, they also have some partial individual moral responsibility for changing Facebook's system; especially as content producers and providers of that system that creates vast profits each year for Facebook through the sale of user's content to advertisers and other information contractors, such as Cambridge Analytica. That responsibility, as Miller correctly argues, is a collective moral responsibility they share jointly.

One way they can do so, is to bring collective civic pressure to bear on Facebook as content producers, to render that system more transparent and accountable by being compliant with the normative requirements and moral responsibilities that Facebook is required to comply with and uphold by virtue of operating as a media company. As we saw in Chap. 4 this also is in keeping with the principles of Stoic philosophy that requires a collective global responsibility and response to the control problem of technology. That is, the unjustified dis-demonic control that the Big Tech companies, such as Facebook and Google, have over our lives and ultimately our well-being. That collective intervention has two aspects.

The first concerns the conflict of roles between Facebook's role as a media company and Facebook's surveillance business model that as we saw earlier in this chapter is conducive to both actual media corruption (as in the case of Cambridge Analytica) and potential corruption in the future. Regarding that concern, the users are entitled to collectively request Facebook to normatively resolve that conflict.

The second concern relates to Facebook's misuse and abuse of its users' privacy and the integrity of its users' information and communications. As producers of Facebook content, which is one of Facebook's most valuable monetary resource, Facebook's users can, therefore, collectively request Facebook to apply more transparent and accountable normative controls and care of their data.

The Systemic Problem of Harvesting and Marketing Information

The problem of the collection and harvesting of users' information via opaque algorithms that is then sold for profit to advertisers, often without the users' knowledge and informed consent, is not limited to Facebook. Insofar as Google, like Facebook, operates as a media company in its information practices through its *search* function and its communication and curations practices that involve the collection, profiling and harvesting of users information via its Apps such as Gmail for example, which like Facebook sells to advertisers for profit; as well as its YouTube practices that have been partly shown to be responsible for the spread of misinformation, fake news and deep fakes, on the Internet generally and specifically during the US presidential elections of 2016;[10] then Google's media role

[10] See, for example, Zeynep Tufekci (2018). YouTube, The Great Radicalizer, *The New York Times*, March 10, 2018; Christopher Mims (2018). Who has more of your personal data than Facebook, The Australian and Wall St Journal, April, 23, 2018; Frank Pasquale (2015) *The Black Box Society: The Secret Algorithms That Control Money and Information*, pp. 50–5 1;66;78–79;96;185;190–191. 214–215; and for a more comprehensive account of the problem, see, Shoshana Zuboff (2019) *The Age of Surveillance Capitalism: The Fight for a Human Future at the New Frontier of Power*; Webb, Amy (2019. Amy Webb (2019) *The Big Nine:*

also places Google like Facebook, within the parameter of the 6th Estate. Therefore, some of the *types* of normative issues examined in this chapter with regard to Facebook, apply equally to Google; for example, its selection, collection and harvesting of users' information via its opaque algorithms, which then sells for profit to advertisers without their knowledge or informed consent. As in the case of Facebook, that problem is systemic: A conflict of interest between the collection and curation of users' information in its role of a media company, and its business model that is specifically designed and geared to generate massive profits through the sale of users' information to advertisers. And as in the case of Facebook, the more time users spend on Google search and more clicks on its other apps and more watch time on YouTube, the more money Alphabet-Google makes. That in itself is not a problem. The problem is the monetisation and weaponisation of the information of its users that gives Google enormous power and control to influence every aspect of our lives.

Operating like Gyges both within and outside the Social Contract for the common good,[11] Google operates under the concealment of its opaque AI-driven algorithms in near-monopolistic manner that allows it to evade and avoid the epistemic and ethical responsibilities of transparency, accountability, truthfulness and trustworthiness by which other media companies are bound. As in the case of Facebook' operations, illustrated by the Cambridge Analytica case, Google's operations undermine citizens' democratic rights to privacy, autonomy and well-being, as well as the democratic system and its associated institutions. The real problem of Google's and Facebook's method and practices is their invasive and non-transparent collection of users' data and its unfair and unaccountable and non-transparent use (Pasquale 2015). The blurring and blending of their information practices with advertising, just as in the case of cash-for-comment and advertorials in some of the practices of the 4th and

How the Tech Titans and Their Thinking Machines Could Warp Humanity, Galloway, Scott). *The Four: The Hidden DNA of Amazon, Apple, Facebook and Google, (2017)*; Van Dijck, Poell and De Waal (2018), *The Platform Society: Public Values in a Connected World* and Lynch, Michael P. (2016) The Internet of Us: Knowing and Understanding Less in the Age of Big Data.

[11] The Social Contract referred to here is based on the philosophical ethical theories of Thomas Hobbes, John Locke and John Rawls that will be examined in more details in Chap. 7. Although different in their approach and articulation those social contract theories counsels every citizen and corporations to act for the common good of society as whole and not for their own exclusive self-interests in violation of the common good.

5th Estates, becomes a much bigger and uncontrollable epistemic, ethical and eudaimonic problem when it's done at the unprecedented stratospheric global level and scale in which both Google and Facebook now operate. As Frank Pasquale poignantly puts it in regard to Google, but equally applicable to Facebook, "Don't Be Evil" is a thing of the past; you can't form a trusting relationship with a black box" (Pasquale 2015). Referring to the Myth of Gyges on which the *model of media corruption* presented in this chapter is based, Frank Pasquale correctly observes:

> Black box insiders are protected as if they are wearing a Ring of Gyges – which grants its wearer invisibility but, Plato warns us in *The Republic*, is also an open invitation to bad behavior. (2015)

He goes on to ask the very crucial question that philosophers since Plato have been asking, and a question Shoshana Zuboff also asks in the her book *The Age of Surveillance Capitalism* (2019) and I am now asking in this book, "what kind of society does this create"? Is it one we want and is it fit for human beings? A society in which the primary purpose of most human beings, as the Stoic and other Hellenistic philosophers and before them Plato and Aristotle clearly identified, is a good and worthy life that leads to the attainment of eudaimonia or well-being? Can we be happy in a world dominated and ruled by a handful of Tech Oligarchs, accountable only to themselves and pursuing their own self-interest over that of the common good of society? A world not of our making or design that is neither within our control, nor one that we have chosen. As Frank Pasquale astutely observes (2015),

> Accountability requires human judgement, and only humans can perform the critical function of making sure that, as our social relations become ever more automated, domination and discrimination aren't built invisibly into the [black boxes] code.

The major systemic problem identified in this chapter at the heart of Facebook's operations and analogously at the core of Google's operations, is an inherent conflict of interests between, on the one hand, Facebook's role as a media company and the normative responsibilities and commitments it has for the safeguarding of its users' information, as required by the Dual Obligation Information Theory (DOIT); and on the other hand, Facebook's financial interests that are determined by an invasive and aggressive business model of maximizing profits through the sale of users' data to advertisers; a business model that is by design geared to override and exert priority for Facebook's financial interests over the users' rights over their information.

Insofar as a similar systemic problem is evident in Google's operations identified also as conflict of interest between its business model that favours Google's financial interests over the normative informational rights of it users, then Google's systemic problem, just as in the case of Facebook, as analysed in this chapter, is conducive to systemic media corruption. Systemic institutional corruption is, therefore, particularly problematic, and harmful to society overall. That problem is analogous and by extension applicable to the Stoic problem of control applied to the problem of control of technology we examined in Chap. 2. Namely, who is in control of our autonomous purposive agency that is an underlying necessary condition for our well-being? Us or the Big Tech companies, through the use of their opaque persuasive algorithms that control most, if not all, aspects of our lives? If as argued in Chap. 3, it is the former, then we should protect and safeguard our autonomous purposive agency as both our individual and collective well-being is dependent upon it.

In a connected digital age, Stoic philosophy, as we saw in Chap. 3, offers a systemic connective practical philosophy of human flourishing and well-being, that combines epistemic, ethical and eudaimonic principles clustered around the central concept of Wisdom, that is relevant and able to respond and motivate a theoretical and practical solution to the control problem of Technology.

In keeping with the Stoic principles of living virtuously, rationally, in accordance with Nature, for our individual and collective well-being, as cosmopolitan citizens with a cosmic consciousness for the attainment of well-being, we have, moreover, a collective responsibility, to respond to the problem of Big Tech's monopolistic control over the Digital Environment of the Internet, which has an overall Dis-demonic impact over our lives. That solution would require a multi-disciplinary and collective societal approach, which we will examine in more detail in the following Chaps. 6, 7, and 8.

Conclusion

The overall objective of this chapter has been to demonstrate how some of the information and communication practices of Facebook, constitute media corruption (Spence and Van Heekeren 2005; Spence 2009, 2017,

2020). To that end, the paper examined what the professional and business role of Facebook is regarding its information/communication practices. It then demonstrated that its role of curating and mediating the information and communication provided by its users, which affords them the collective sharing and exchange of knowledge-relevant information on its platform, reveals that Facebook is essentially a *media company*, and not merely a "platform." As such, Facebook is liable to the same normative responsibilities of other media companies based on the Dual Obligation Information Theory (DOIT)-Wisdom model we examined in Chap. 2. As such the DOIT-Wisdom model commits Facebook to a *normative editing responsibility*, qua media company, to ensure both the preventive detection and corrective editing of fake news, as well as other forms of misinformation disseminated on its platform. Based on DOIT's normative requirements, Facebook is furthermore normatively accountable for the harvesting of users' information, which it collects with the aid of algorithms and sells to advertisers and other information brokers for profit, without the users' informed consent. This general practice involves a *conflict of interest* between its role as a media company and its selling of users' information to advertisers for profit without the users' informed consent. That conflict of interest lies at the heart of Facebook's operations primarily as a result of its Business Model and thus, conducive to systemic media corruption.

That type of media corruption referred to in the paper as *advert-facetorial*, is a type of *advertorial*, also present in legacy media. In the case of traditional advertorials by legacy media companies, the content is generated by those companies in partnership with the advertisers and presented to the users as credible and objective independent journalistic content, without the users' awareness of the deception and manipulation. In the case of Facebook, this process is reversed. The *advert-facetorials* are created through users' content without their informed consent, and then sold to advertisers, who in turn, target the users with advertisements, tailored specifically to the users' individual profiles created by Facebook's algorithms. Although the process involving traditional advertorials in the legacy media, and Facebook's surveillance advertising practice based on its business model are different in type, they are similar in kind. Both are designed to deceive, manipulate and exploit their users, through misinformation. Another characteristic these advertorials share in common is a *conflict of interest*. A conflict between a media company's professional role morality of disseminating and communicating information to the public, and in Facebook's case, its own role morality as a media company of mediating and curating its users' information.

Such conflicts of interest involving a conflict of roles, expressed either in the form of advertorials, or in the form of *advert-facetorials* in the case of Facebook, are conducive to media corruption and violate the epistemic, ethical and eudaimonic principles required by the DOIT-Wisdom model. This clearly demonstrates the close conceptual and normative relationship between DOIT-Wisdom and the Media Corruption model used in this chapter, and further demonstrates and explains why media corruption and corruption generally, is *fundamentally a moral problem* with a negative eudaimonic impact on society.

This chapter demonstrated that Facebook's normative violations with regard to DOIT-Wisdom and its related violations with regard to the Media Corruption model as examined in this chapter, constitute media corruption and more generally, ongoing *systemic media corruption*. A corruption that manifest itself through its corrupting effects in at least three ways: As *information and communication corruption*, *political corruption* and because of its corrupting effects on its users qua citizens, *civil corruption*.

The chapter also demonstrated that the problem of the collection and harvesting of users' information via opaque algorithms that is then sold for profit to advertisers, often without the users' knowledge and informed consent, is a *generic systemic problem* not limited to Facebook. Insofar as Google, like Facebook, operates as a media company in its information practices through its *search* function and its communication and curations practices that involve the collection, profiling and harvesting of users information via its Apps such as Gmail for example, which like Facebook sells to advertisers for profit; as well as its YouTube practices that have been partly shown to be responsible for the spread of misinformation, fake news and deep fakes, on the Internet generally and specifically during the US presidential elections of 2016;[12] then Google's media role also places Google like Facebook, within the parameter of the 6th Estate. Insofar as a similar systemic problem is evident in Google's operations identified also as con-

[12] See, for example, Zeynep Tufekci (2018). YouTube, The Great Radicalizer, *The New York Times*, March 10, 2018; Christopher Mims (2018). Who has more of your personal data than Facebook, The Australian and Wall St Journal, April, 23, 2018; Frank Pasquale (2015) *The Black Box Society: The Secret Algorithms That Control Money and Information*, Cambridge, pp. 50–51;66;78–79;96;185;190–191. 214–215; and for a comprehensive account of the problem, see, Shoshana Zuboff (2019) *The Age of Surveillance Capitalism: The Fight for a Human Future at the New Frontier of Power*.

flict of interest between its business model that favours Google's financial interests over the normative informational rights of it users, then Google's systemic problem, just as in the case of Facebook, as analysed in this chapter, is conducive to systemic media corruption. Systemic institutional corruption, as we saw above, is particularly problematic and harmful to the whole of society as it undermines the collective common good of society and its underlying democratic system and aligned institutions. Finally, the systemic problem of media corruption illustrated by the practices of Big Tech companies such as Facebook and Google, is analogous and by extension applicable to the problem of control of technology examined in Chap. 3. Namely, who is in control of our autonomous purposive agency that is an underlying necessary condition for our well-being? Is it us qua citizens, or the Big Tech companies, through the use of opaque persuasive algorithms that exercise informational control over most if not all aspects of our lives? If as argued in Chap. 3, it is the former, then we should protect and safeguard our autonomous purposive agency as both our individual and collective well-being as global citizens is dependent upon it.

References

Ceva, Emanuela and Ferretti, Maria P.(2017). "Political Corruption". *Philosophy Compass,* 12(2): 1–10.

Elliott, Deni. and Spence, Edward. (2018) *Ethics for a Digital Era*. Oxford: Wiley-Blackwell.

Galloway, Scott. (2017). *The Four: The Hidden DNA of Amazon, Apple, Facebook and Google*. London, Great Britain: Bantam Press.

Gewirth, Alan. (1978). *Reason and Morality*, Chicago, University of Chicago Press.

Lessig, Lawrence. (2013). "Institutional Corruption - Defined". *Social Sciences,* 41(3):553.

Milkman, Oliver, 2019. "Defiant Mark Zuckerberg Defends Facebook Policy to Allow False Ads." *The Guardian*. December 2, 2019.

Miller, Seamus. (2010). *The Moral Foundations of Social Institutions*. Cambridge University Press.

Miller, Seumas, (2018). "Corruption". *Stanford Encyclopedia of Philosophy*. (Winter 2018 Edition), Edward N. Zalta (ed.).

Miller, Seumas., Roberts, Peter. Spence, Edward. (2005). *Corruption and Anti-corruption: An Applied Philosophical Approach*. Upper Saddle River, NJ, USA: Pearson/Prentice Hall.

Mims, Christopher (2018). "Who Has More of Your Personal Data Than Facebook?" *The Australian and Wall St Journal*, April 23, 2018.

Pasquale, Frank. (2015) *The Black Box Society: The Secret Algorithms That Control Money and Information.* Cambridge, Massachusetts: Harvard University Press
Plato. (2007). *The Republic.* Penguin Classics.
Spence, Edward. (2018). "Harm in Media Marketing: The Branding of Values", in Patrick Plaisance (ed.), *Communication and Media Ethics*, Berlin, Boston: Walter de Gruyter Mouton.
Spence, Edward. (2017). "Corruption in the Media," in Michael S. Aßländer and Sarah Hudson (eds.), *The Handbook of Business and Corruption: Cross-Sectoral Experiences*, Emerald Group Publishing, 453–480.
Spence, Edward. (2014). "The Advertising of Happiness and the Branding of Values." In Michael Boylan (Ed.). *Business Ethics* (2nd edition). Upper Saddle River, NJ, USA: Pearson/Prentice Hall.
Spence, Edward and Van Heekeren, Brett. (2005). *Advertising Ethics,* New Jersey, USA: Pearson, Prentice Hall.
Spence, Edward, Alexandra, Andrew, Quinn, Aaron, and Dunn, Anne. (2011). *Media, Markets and Morals.* Oxford: Wiley-Blackwell.
Spence, E. (2011) "Information, Knowledge and Wisdom: Groundwork for the Normative Evaluation of Digital Information and Its Relation to the Good Life," *Ethics and Information Technology*, Volume 13, Number, 3, Pages 261–275.
Spence E. (2009) "A Universal Model for the Normative Evaluation of Internet Information," *Ethics and Information Technology*, Volume, 11, Number 4, Pages 243–253.
Spence, Edward (2008). "Corruption in the Media." *International Journal of Applied Philosophy*, Volume 22:2 – Fall 2008, pp. 231–241.
Spence, E. *Ethics Within Reason: A Neo-Gewithian Approach* (2006). Maryland: Lexington Books (a division of Rowman and Littlefield, USA).
Spence, Edward (2006a) "Corruption Doomed to Fail in Long term." *Financial Review*, January 14, 2006.
Tufekci, Zeynep. (2018a). "Facebook's Surveillance Machine", *The New York Times*, March 19, 2018.
Tufekci, Zeynep (2018b). "YouTube, The Great Radicalizer", The *New York Times*, March 10, 2018.
Zuboff, Shoshana. (2019). *The Age of Surveillance Capitalism: The Fights for A Human Future At The New Frontier of Power.* London: Profile Books Ltd.
Zuckerberg, Mark. (2019). "Four Ideas to Regulate the Internet", *Facebook*, March 30, 2019.
Zuckerberg, Mark. (2020). "Big Tech Needs More Regulation", *Financial Times*, February 18, 2020.

CHAPTER 6

Wisdom and Intelligent Machines: Who Is in Control?

The ultimate form of control is individual sovereignty, understood as self-ownership, especially one's own body, choices, and data. The fight for digital sovereignty is an epochal struggle
—Luciano Floridi (2019)

Introduction

This chapter undertakes a critical examination and analysis of the concept of dignity, a central tenet of both Stoic and neo-Stoic philosophy, in relation to the problem of control. Specifically, it explores the application of this philosophical concept to technology, with a particular focus on Information and Communication Technologies (ICTs) and Artificial Intelligence (AI). Given that many critical issues in the use of these technologies arise from the deployment of AI algorithms, this chapter addresses these issues alongside the normative impact of such technologies on individual and societal well-being. A detailed evaluation will be conducted on the implications of ICTs and AI technologies for well-being, focusing on two prominent challenges in AI: the inscrutability or "black box" problem, and the control problem.

In addressing these challenges, the chapter will propose potential solutions, including some recently advanced by AI researchers. A central concern is that the lack of transparency in AI systems undermines accountability, and without accountability, there exists no effective normative framework

E. H. Spence, *Wisdom in the Age of Intelligent Machines*,
https://doi.org/10.1007/978-3-031-93883-2_6

to monitor or control these systems to mitigate potential adverse societal impacts. Consequently, the collective good of society and humanity at large is at risk.

The chapter will also introduce the concept of anticipatory normative controls through design, aimed at preventing or minimising the risks posed by the opacity and lack of control in AI technologies. Furthermore, as discussed in Chap. 3, the overarching question guiding this analysis—and indeed the broader discourse of this book—is: who ultimately controls technology—society or the Big Tech corporations?

The Control Problem and Digital Sovereignty

Luciano Floridi defines sovereignty as "a form of legitimate, controlling power" (2019) and elaborates on "control" in relation to sovereignty as

> [t]he ability to influence something (e.g., its occurrence, creation, or destruction) and its dynamics (e.g., its behaviour, development, operations, interactions), including the ability to check and correct for any deviation from such influence. In this sense, control comes in degrees and above all can be both pooled and transferred.

This notion of control is central to understanding sovereignty, particularly in the context of individual sovereignty, which Floridi describes as self-ownership, encompassing one's body, choices and data. Floridi emphasises that the "fight for digital sovereignty is an epochal struggle," characterised by an asymmetric power dynamic between corporations and states.

The key terms in Floridi's definition resonate strongly with Stoic philosophy, particularly regarding the control problem of technology discussed in previous chapters, especially Chap. 7. As established in Chaps. 3 and 7, Stoic philosophy holds that the ultimate and exclusive form of control human beings possess lies in their judgements, choices and impulses. Collectively, these constitute what the Stoics refer to as *prohairesis* or *hegemonikon*, the faculty of inner volition. Importantly, the Stoics argue that this inner control does not extend to external circumstances, including the body, which is subject to external contingencies beyond one's control.

From a Stoic perspective, individual sovereignty—analogous to autonomy—faces significant challenges in the digital age. As explored in Chap. 3,

Big Tech corporations exploit their asymmetric power through persuasive technologies that infiltrate and manipulate individual autonomy, undermining both personal sovereignty and collective well-being. This dynamic not only compromises individual autonomy but also fosters systemic media corruption, as illustrated by cases such as Cambridge Analytica.

The monopolistic hegemony of corporations like Facebook and Google exacerbates this problem by eroding individual sovereignty, which the Stoics identify with the inviolable *hegemonikon*. Floridi astutely highlights that the digital age compels a rethinking of sovereignty, noting:

> Modern analogue sovereignty is still necessary but increasingly insufficient … Once cumulative advantages create a monopolistic regime, then there is no real competition, so no real consumer's choice, and therefore no real accountability of the companies that dominate specific markets.

This monopolistic regime transforms individual citizens into "follower-users" under the influence of what Zuboff (2019) terms "Surveillance Capitalism." Floridi further observes that profiles and data, created and controlled by multinational corporations, epitomise the globalised nature of this issue, necessitating urgent reconsideration of sovereignty in the digital age.

The pressing question of who is best positioned to safeguard individual and collective digital sovereignty against the normative abuses of Big Tech monopolies will be addressed in greater depth later in this chapter. For now, it is worth noting that several legal actions targeting anti-competitive practices by Big Tech corporations are already underway as of December 2020, including in the USA, EU and Australia. The July 29, 2020, hearing of the House Judiciary Subcommittee on Antitrust, Commercial and Administrative Law, which involved the CEOs of Amazon, Apple, Google and Facebook, marks a significant moment in the broader struggle to regulate Big Tech's dominance.

From a Stoic and neo-Stoic perspective, the current monopolistic control exercised by Big Tech over digital data represents a profound challenge to the legitimacy of such control. This issue transcends external contingencies, directly impacting what ought to remain within individual control—autonomy, sovereignty and dignity. As Floridi rightly notes, the illegitimacy of this control strikes at the heart of individual sovereignty, which, is grounded in natural rights to freedom and well-being. These rights are foundational and inviolable, rooted in the dignity of individuals as autonomous agents.

The Control Problem and the Normativity of Algorithms

Building on the previous analysis, this section examines the normative issues raised by algorithms and their impact on rights, autonomy and sovereignty. Specifically, it explores how algorithms undermine personal and collective autonomy by eroding control over decision-making processes. This challenge arises from the scale, complexity and opacity of AI-driven decision-making, which is often executed autonomously and lacks transparency, leading to a loss of human oversight. Consequently, the removal of autonomy through machine learning algorithms equates to the removal of human control, as autonomy defines and determines both personal and collective agency (Mittelstadt et al. 2016).

As Mittelstadt et al. (2016) highlight:

The algorithm defines decision making rules to handle new inputs. Critically, the human operator does not need to understand the rationale of decision-making rules produced by the algorithm… learning capacities grant algorithms some degree of autonomy. The impact of this autonomy must remain uncertain [emphasis added] to some degree. As a result, tasks performed by machine learning are difficult to predict beforehand… Uncertainty can thus inhibit the identification and redress of ethical challenges in the design and operation of algorithms.

This dual loss of control, both internal and external, is significant. Internally, even Big Tech companies experience diminished oversight as algorithms evolve beyond their operators' comprehension. Externally, regulatory bodies and users face reduced ability to govern or counteract the influence of these systems. Algorithms developed by companies such as Facebook and Google are designed to influence user behaviour, undermining autonomy and decision-making for their self-interest. This systematic erosion of user autonomy and control constitutes a violation of fundamental rights, including freedom, well-being and dignity.

The infringement on dignity, as identified in this chapter is an absolute human right and is particularly concerning. By targeting users' decision-making and autonomy through persuasive and invasive algorithms, Big Tech companies breach ethical principles, prioritising corporate gain over individual sovereignty. To better understand these ethical challenges, Mittelstadt et al. (2016) provide a framework, categorising the ethics of algorithms into six distinct areas:

1) Inconclusive evidence
2) Inscrutable evidence
3) Misguided evidence
4) Unfair outcomes leading to discrimination
5) Transformative effects
6) Traceability

Inconclusive evidence occurs where the algorithms produce probability, yet inevitably uncertain knowledge, as the statistical methods employed can identify significant correlations but insufficient to indicate a causal connection. This is clearly an epistemic limitation but also an ethical concern as the authors themselves conclude, "the risk of being wrong affects one's epistemic responsibilities"; as a wrongly interpreted correlation may result in harm, as for example, when used in police profiling, when a subject is unjustly statistically correlated to a crime they may not have committed. The main epistemic and ethical concerns under the criterion of inconclusive evidence, is that (a) "causality is not established prior to acting upon the evidence produced by the algorithm" and (b) an issue of a crucial methodological concern is that even if, as the authors correctly point out, the correlation or causal knowledge is strong "this knowledge may only concern populations while actions are directed towards individuals." They go on to correctly point out that "actions taken on the basis of inductive correlations [at the population-level] have real impact on human interests independent of their validity" (2016). Is this a case of ethics by numbers? This is a crucial question we shall pursue later as it impacts on the very notion of individual autonomy, which is at the heart of most of the issues we will examine in this chapter.

Inscrutable evidence represents a major normative issue, particularly in its implications for transparency, control and accountability. As previously discussed, the opacity of algorithms undermines accountability and control, thereby eroding oversight. This opacity enables Big Tech companies to consolidate control and exercise dominance over both individual and collective data. Mittelstadt et al. (2016) identify two primary components of transparency: accessibility and comprehensibility of information. Both are systematically denied by Big Tech, whose deliberate secrecy surrounding the design and application of algorithms perpetuates a significant asymmetry of information. This asymmetry privileges the commercial and financial interests of these corporations over the rights and interests of users and the broader public good.

The asymmetry of information, in turn, creates an imbalance in knowledge and decision-making power, granting Big Tech companies a decisive advantage as curators, possessors and processors of digital data. This dynamic undermines collective normative goods—epistemically, ethically and eudaimonically—both for individuals and for society. As Mittelstadt et al. rightly note, "the opacity of machine learning algorithms inhibits oversight" (2016). Without oversight, there can be no meaningful transparency, accountability, or control. The authors further emphasise that for transparency to be effective, information must not only be accessible but also comprehensible: "Besides being accessible, information must be comprehensible to be transparent" (2016). This point is crucial. Data subjects have a legitimate interest in understanding how information collected about them is used and how algorithmic decisions are made. Without such understanding, the asymmetry of information constitutes a violation of users' autonomy and sovereignty over their digital data.

As Mittelstadt et al. observe, oversight and human intervention in algorithmic decision-making are severely constrained by the inherent complexity and speed of machine learning algorithms:

> Meaningful oversight and human intervention in algorithmic decision making is impossible when the machine has an information advantage over the operator… [or] when the machine cannot be controlled by a human in real-time due to its processing speed and the multitude of operational variables. (Matthias 2004)

This "black-box problem," caused by the opacity and incomprehensibility of algorithmic operations, restricts oversight both internally (by operators) and externally (by governments, regulatory institutions and society). Furthermore, algorithms' inscrutability makes their decisions difficult to understand and challenge, especially given the existing asymmetry of information. This asymmetry unfairly favours Big Tech, undermining the interests of users and society at large. Mittelstadt et al. succinctly summarise the normative stakes:

> Under these conditions, decision-making is poorly transparent… and the failure to render the processing logic comprehensible to data subjects disrespects their agency… meaningful consent to data processing is not possible when opacity precludes risk assessment. (2016)

The emphasis on respect and risk assessment is critical to evaluating the normativity of information and communication technologies (ICTs) and AI systems. As explored in earlier chapters, respect for autonomy, dignity and self-determination are fundamental rights underpinned by Alan Gewirth's Principle of Generic Consistency (PGC) and the DOIT-Wisdom model, as well as the traditions of Stoic and neo-Stoic philosophy. Risk assessment, similarly, is essential for legislative and societal control over Big Tech's informational and surveillance practices, which systematically undermine these principles. Stoic philosophy underscores the importance of mitigating risks by focusing on factors within our control, such as autonomous volition (hegemonikon). Yet, as demonstrated, our autonomy, dignity and freedom are being systematically eroded by opaque technologies. The unchecked power of Big Tech poses an existential risk, demanding that their illegitimacy be exposed and brought under societal control to safeguard the common good.

As Shoshana Zuboff aptly notes, "we have become the investment futures in a surveillance culture dominated by the Big Tech companies" (Orlowski 2019). This surveillance-driven culture further exacerbates trust issues between data subjects and Big Tech companies. Transparency is foundational to trust, as Mittelstadt et al. observe:

> Transparency disclosures by data processors and controllers [such as Big Tech companies] may prove crucial in the future to maintain a trusting relationship with data subjects. (2016)

Onora O'Neill's (2013) distinction between trust and trustworthiness, as outlined in her 2013 TED Talk, provides a useful framework for evaluating Big Tech's practices. According to O'Neil, trustworthiness requires competence, honesty and reliability—all of which must coexist. While Big Tech's algorithms demonstrate technical competence, they fall short on honesty and reliability due to their lack of transparency and accountability, as well as the informational asymmetry they perpetuate. These failures undermine trustworthiness and highlight the manipulative intent embedded in their business models.

This problem becomes even more critical in the context of future developments in AI, particularly the potential emergence of superintelligent systems whose decision-making may far exceed human comprehension. The implications of such advancements will be explored further in Chap. 7. For now, it is evident that the opacity of algorithms and the asymmetry of information they perpetuate are not merely technical challenges but profound ethical and existential issues demanding urgent scrutiny and reform.

Misguided Evidence and Unfair Outcomes Leading to Discrimination

I have grouped these two issues together as they are closely interrelated. As Mittelstadt et al. (2016) observe:

> Algorithms inevitably make biased decisions … As a result, the values of the author [of an algorithm], wittingly or not, are frozen into the code, effectively institutionalising those values … social biases can be embedded in system design purposefully by individual designers, seen for instance in manual adjustments to search engine indexes and ranking criteria … machine learning algorithms trained from human-tagged data inadvertently learn to reflect biases of the taggers … Flaws in the data are inadvertently adopted by the algorithms and hidden in outputs and models produced … [and] algorithms also require interpretation [and] explaining the correlation requires additional justification.

This analysis of misguided evidence challenges the assumption that algorithms and their use in making statistical correlations—which are then applied to individuals, often with unjust and harmful consequences—are objective or free from bias. At least with human biases, there is the possibility of investigation, exposure and public challenge. By contrast, autonomous algorithms obscure such biases, making them harder to detect and critique. This should give us serious pause when considering the uncritical use of algorithms without human oversight and control.

Mittelstadt et al. (2016) further emphasise that "much of the reviewed literature addresses how discrimination results from biased evidence and decision-making." Profiling by algorithms is a primary cause, as these systems identify correlations and make predictions about behaviour at a group level rather than focusing on actual individual behaviour. Such practices inadvertently create evidence bases that reinforce discrimination. They note:

> Bias is a dimension of the decision-making itself, whereas discrimination describes the effects of a decision, in terms of adverse disproportionate impact resulting from algorithmic decision-making … [and] the capacity of individuals to investigate the personal relevance of factors used in decision-making is inhibited by opacity and automation.

The issues we have explored so far share a common thread: the use of algorithms, particularly by Big Tech companies, undermines human autonomy and freedom. Algorithms increasingly intrude into the one core area that the Stoics identified as central to human control—our capacity for rational and virtuous judgements and choices. Now, even this internal sphere of autonomy is accessed, assessed and manipulated by inscrutable and unaccountable algorithms relying on statistical correlations at the group level. These processes have profound epistemic, ethical and eudaimonic consequences, creating a society dangerously close to the dystopian vision outlined in Orwell's *1984*.

Unlike Orwell's nightmarish vision, however, there is still time to challenge this socially unhealthy and unacceptable trajectory. We must hold Big Tech companies accountable and demand greater transparency and oversight before it is too late. Our collective future depends on it.

Transformative Effects of Algorithmic Personalisation

Autonomy and Persuasive Technologies

A significant concern under the topic of transformative effects is the threat posed by algorithmic personalisation to individual autonomy. This issue has been critically examined through Stoic and neo-Stoic philosophy in Chap. 3, where technology's capacity to influence judgements and decision-making is scrutinised. Big Tech companies design persuasive technologies with the express purpose of exerting control over the autonomy of data subjects. These technologies leverage user data to nudge decisions in ways that prioritise corporate interests over individual preferences, ultimately undermining autonomy.

As Mittelstadt et al. (2016) observe:

> Personalization algorithms tread a fine line between supporting and controlling decisions by filtering which information is presented to the user based upon in-depth understanding of preferences, behaviours, and perhaps vulnerabilities to influence. … The subject's autonomy in decision-making is disrespected when the desired choice reflects third-party interests above the individual's.

These observations underscore how personalisation algorithms filter information to shape decisions, often coercing users into "institutionally preferred actions" (Johnson 2014). For instance, consumers can be nudged toward market needs through the display and ranking of products. The idea of teaching algorithms to balance coercion and support for user autonomy, as Mittelstadt et al. suggest, is theoretically appealing but practically unfeasible due to the business models of Big Tech companies, which are fundamentally oriented towards influencing user autonomy for profit.

Mittelstadt et al. further point out:

> Filtering algorithms [through personalisation] that create "echo chambers" devoid of contradictory information may impede decisional autonomy. Algorithms may be unable to replicate the "spontaneous discovery of new things, ideas and options" which appear as anomalies against the subject's profiled interests.

This lack of control over one's data autonomy, exacerbated by informational asymmetries between Big Tech and users, fundamentally undermines autonomy as conceptualised by Stoic philosophy. For the Stoics, control over one's judgements and choices—prohairesis—is central to well-being (eudaimonia). The external control exerted by algorithms thus jeopardises this essential autonomy, rendering individuals vulnerable to manipulation.

Privacy and Informational Identity

Another critical issue is the transformation of privacy by algorithms. As Mittelstadt et al. (2016) note:

> Algorithms are also driving a transformation of notions of privacy. … A right to identity derived from informational privacy interests suggests that opaque or secretive profiling is problematic.

Carissa Véliz (2020) highlights the societal consequences of surveillance capitalism in *Privacy is Power*:

> Surveillance threatens freedom, equality, democracy, autonomy, creativity, and intimacy. … Having too little privacy is at odds with having well-functioning societies. Surveillance capitalism needs to go. … Privacy is a right for good reasons, defend it.

For the Stoics, as for Aristotle, the violation of privacy undermines autonomy and dignity. Informational privacy is crucial for maintaining self-sovereignty over judgements and choices. Allowing third parties to access and control personal information without informed consent constitutes a fundamental violation of autonomy. As previously discussed, this external interference compromises the Stoic ideal of self-determined action and, by extension, one's eudaimonia.

Mittelstadt et al. also emphasise the broader implications of profiling:

> Identity is increasingly influenced by knowledge produced through analytics that makes sense of growing streams of behavioural data. Profiling seeks to assemble individuals into meaningful groups, for which identity is irrelevant. … External identity constructions by algorithms create de-individualization… judging and treating people on the basis of group characteristics instead of on their own individual characteristics and merit.

The reliance on group-level profiling to make decisions that impact individual lives introduces a paradoxical conflict between digital group identities and natural individual identities. Decisions based on algorithmic profiles can result in unjust outcomes for individuals whose natural identities are subordinated to statistical group characteristics.

The Borg Metaphor: Digital vs. Natural Identity

The division between digital group identity (DGI) and natural individual identity (NII) can be likened to the Borg in *Star Trek*. In this fictional universe, individuals assimilated into the Borg collective lose their personal identities, becoming part of a hive mind. This metaphor illustrates the existential risks posed by algorithms that construct and control digital identities. Like the Borg's victims, individuals may find their autonomy subsumed by externally imposed digital identities that conflict with their natural selves.

As illustrated by the character Seven-of-Nine in *Star Trek: Voyager*, reclaiming one's individuality after assimilation is arduous and fraught with challenges. Similarly, escaping the systemic control of digital identities constructed by Big Tech's persuasive technologies demands significant effort and resistance. Yet, as the Borg warn, "Resistance is futile."

Normative Implications and Existential Challenges

The observations made by Mittelstadt and others highlight the normative tensions between autonomy, privacy, identity and control. Algorithms' capacity to digitalise and datafy identity introduces profound challenges to individual freedom, dignity and well-being. The external control exercised by Big Tech over digital identities undermines the internal control individuals should have over their natural identities. This systemic imbalance demands a re-evaluation of the ethical frameworks governing algorithmic personalisation. In conclusion, the threat to autonomy, privacy and identity posed by algorithmic personalisation necessitates urgent philosophical and practical responses. The Stoic emphasis on self-control and dignity offers a compelling lens through which to critique these transformative effects and advocate for the restoration of autonomy in the digital age.

Traceability

In this final section, Mittelstadt et al. examine the traceability of algorithms and the moral responsibility attributable to them (Mittelstadt et al. 2016). As with the preceding sections, this discussion is framed within the broader normative issue of the control problem of technology, specifically in relation to the autonomy of individuals as data subjects in their dual capacity as both agents and persons.

This section raises several critical concerns: Who bears normative responsibility for the decision-making processes of algorithms, given their inherent opacity and lack of transparency? Should autonomous or semi-autonomous machine learning algorithms be held wholly or partially accountable for their decisions? Who should oversee their accountability, and how should such oversight be implemented? Furthermore, should machine learning algorithms be designed to incorporate ethical reasoning and behaviour? Another key issue concerns the relationship between group-level digital identity (DGI) and natural individual identity (NII). If both identities are, to some extent, controlled by the algorithms of Big Tech companies, what are the normative implications for personal autonomy—both digital and natural? Moreover, who should ensure that individuals' rights as agents and persons are safeguarded against potential manipulation and exploitation by these companies?

Mittelstadt et al. highlight several interrelated concerns that warrant careful examination:

- Ethical responsibility for algorithms—who is ultimately responsible?
- Human oversight and accountability—how should it be implemented, and by whom?
- The challenge of designer control versus the autonomous behaviour of machine learning algorithms, which creates an accountability gap.
- Whether autonomous algorithms should bear ethical responsibility.
- The extent to which moral responsibility should be distributed between human operators and algorithms.
- The feasibility and implications of embedding moral reasoning and behaviour into AI systems.

As reiterated throughout this chapter, the overarching theoretical and practical concern underlying these issues is the normative problem of control. Specifically, who should govern autonomous algorithms and AI agents—individuals, collectives, or Big Tech corporations? As explored in this chapter and in Chap. 3 through the lens of Stoic and neo-Stoic philosophy, the control problem directly pertains to human autonomy and well-being. Autonomy is fundamentally tied to an individual's capacity for self-governance over their judgements, decisions and choices. However, as previously examined, the informational asymmetry between individuals and Big Tech companies, coupled with the deliberate targeting and erosion of autonomy through persuasive technologies and algorithmic manipulation, places both individual and collective self-governance at risk.

Building on these concerns, the following section will extend the discussion of control and autonomy by addressing what I refer to as the **identity paradox**—an implicit yet significant dimension of the control problem.

The Identity-Paradox

The Identity-Paradox and the Normative Implications of Algorithmic Control

The identity-paradox, a critical issue previously explored in the subsection *Transformative Effects*, concerns the existential and normative challenges posed by algorithmic digitalisation and datafication of individual identity. This process is driven by opaque, incomprehensible and unaccountable algorithms that construct digital identities for individuals without their knowledge or consent. These digital identities are derived from statistical group-level data, which often bears little to no resemblance to an individual's natural identity. Despite this discrepancy, decisions impacting an individual's real-life circumstances are increasingly based on these algorithmically generated profiles, leading to potentially unjust outcomes, particularly through impersonal and automated profiling (see subsection *Unfair Outcomes Leading to Discrimination*).

A particularly troubling consequence of this phenomenon is the potential for digital identity to be placed in normative opposition to natural identity. For example, algorithmic profiling may result in individuals being categorised as potential criminals based on statistical correlations within their group-level digital identity, even though their natural identity remains entirely unaware of this classification. This raises significant ethical concerns regarding due process, fairness and the presumption of innocence.

Beyond these immediate concerns, an even more insidious existential threat emerges: the fragmentation of identity into two distinct entities—an externally controlled group-level digital identity and an internally governed natural individual identity. This division is further exacerbated by the pervasive influence of persuasive technologies deployed by Big Tech companies such as Facebook and Google. These technologies, designed to manipulate user behaviour and decision-making, undermine individual autonomy by exerting systemic control over both digital and natural identities, often without the individual's awareness or informed consent.

From a Stoic and neo-Stoic perspective, this level of external control over identity is normatively unacceptable as it constitutes a violation of fundamental autonomy. As examined, individuals possess prima facie rights as agents and absolute rights as persons, both of which are essential for self-determination. The dual-aspect theory of personal identity and agency, supported by Kantian and Gewirthian rational arguments, alongside Stoic and neo-Stoic principles, allows for a robust critique of this identity-paradox. Specifically, the algorithms employed by Big Tech corporations simultaneously erode both individual autonomy and digital autonomy, resulting in a twofold violation of dignity—first, as digital agents subjected to opaque algorithmic classification, and second, as individual persons whose autonomy is systematically compromised.

Ethical and Normative Consequences

The normative harm inflicted upon data subjects—and by extension, society as a collective of individual persons and citizens—is therefore twofold. This brings into focus the question of what ethical and regulatory measures must be taken to either eliminate or mitigate such harms. Mittelstadt et al. (2016) argue that "automation of decision-making creates problems of ethical consistency between humans and algorithms." This concern aligns with the Stoic principle that consistency in virtuous conduct is essential for maintaining personal integrity—an integral component of normative autonomy and *hegemonikon* (volition). Integrity, in turn, is crucial for making rational decisions that contribute to human flourishing (*eudaimonia*).

The Stoic and neo-Stoic philosophical framework, with its emphasis on rationality, virtue and logical argumentation, provides a critical foundation for addressing these ethical challenges. As discussed in Chap. 3, Gewirthian neo-Stoic philosophy formulates natural rights to freedom and well-being based on the dual concepts of agenthood and personhood. These rights are inherently tied to the value of dignity, as elaborated in Chap. 6 under the section *The Dignity-Conferring Value of Rights.*

A key contribution of the neo-Stoic model presented in this book is its ability to evaluate the normative impact of Big Tech's incursions into personal autonomy through a multifaceted ethical lens. This model, drawing from Gewirth's rationalist ethics of rights (see Chap. 3), facilitates a comprehensive assessment of algorithmic ethics by integrating rationalist ethics, virtue ethics and *eudaimonic* ethics. As analysed in Chap. 2, these three ethical approaches are foundational to the DOIT Wisdom normative model, which provides a structured method for assessing the ethical implications of AI-driven technologies.

Although Mittelstadt et al. highlight the absence of a "consensus view" regarding how to practically redistribute social and ethical duties displaced by automation (2016), this challenge is not insurmountable. A closer examination reveals that most ethical theories—including utilitarianism, contractarianism, communitarianism and ethics of care, as well as rationalist theories such as those of Gewirth and Kant—converge on key normative principles. These include respect for dignity, equal rights, harm mitigation, the promotion of the collective good and duties of care and fairness.

From the perspective of these ethical frameworks, the pervasive and manipulative practices of Big Tech, which systematically erode individual autonomy and societal well-being, must be regarded as normatively problematic. As demonstrated throughout this book, these practices constitute a fundamental violation of epistemic, ethical and *eudaimonic* rights. Furthermore, they undermine not only individual agency but also the collective good of society, reinforcing the urgency of regulatory and philosophical interventions to safeguard personal and civic autonomy.

Shared Responsibility, Compliance and Legislation

Mittelstadt et al. (2016) highlight two critical concerns: (a) tracing ethical responsibility between human operators and autonomous algorithms and (b) ensuring accountability and ethical compliance. They warn that substantial trust in algorithms can lead to "de-responsibilisation" of human actors, who assume automated processes are inherently correct (Zarsky 2016). This delegation of decision-making risks shifting responsibility away from human oversight and justifying harmful outcomes (Davis et al. 2021). Further research is needed to mitigate these risks and establish resilience as a core ethical principle in algorithm design.

The fundamental issue with trusting autonomous algorithms lies in their opacity and the power imbalance between users and Big Tech companies. Given the inscrutability of these systems, uncritical trust is unwarranted. As Mittelstadt et al. observe, de-responsibilisation not only undermines human agency but, as I argue, also contributes to the de facto dehumanisation of society. Autonomy, a fundamental human right emphasised by the Stoics and neo-Stoics (see Chap. 3), must be preserved. Surrendering control over our autonomy—whether voluntarily or through systemic manipulation—compromises our well-being and constitutes a failure of ethical duty. The unchecked influence of opaque, unaccountable algorithms on individual autonomy violates our dignity and fundamental rights.

On transparency and compliance, Mittelstadt et al. acknowledge the unresolved challenge of operationalising transparency in machine learning (2016). Mere code disclosure is insufficient; effective algorithmic auditing is essential. Scholars advocate for both internal audits by tech firms (Zarsky 2016) and external oversight by regulators and researchers (Pasquale 2015; Kitchin 2016). A cooperative approach between policymakers, developers and independent auditors is required to ensure accountability.

Mark Zuckerberg, in his 2020 *Financial Times* op-ed, supports increased regulation, arguing that the ethical integrity of online information should be a collective responsibility shared by companies, governments, media and citizens. Similarly, Van Dijck et al. (2018) stress that platform governance requires an integrated approach to uphold public trust and democratic values. Their call for shared responsibility aligns with my argument in *Media Corruption in the Age of Information* (Spence 2021a) and supports the need for a global regulatory framework. The DOIT-Wisdom theory (see Chap. 2) provides a normative model for ensuring ethical governance of digital platforms. Given the global nature of the Internet, ethical regulation must likewise be universally applicable.

Tech Regulation

Mark Zuckerberg, in his 2020 Financial Times op-ed, highlights a crucial dimension of regulation—namely, oversight and accountability that extends beyond self-regulation to include clear, external government intervention. As he states,

> People need to feel that global technology platforms answer to someone, so regulation should hold companies accountable when they make mistakes.

Further, he argues:

> If we don't create standards that people feel are legitimate, they won't trust institutions or technology. (Zuckerberg 2020)

This statement is significant for two key reasons. First, it underscores not only retrospective regulation—such as the substantial fines imposed on companies like Facebook and Google for past violations—but also prospective regulation through ongoing oversight. The term "oversight" suggests a proactive regulatory framework, akin to the annual independent auditing of financial statements and tax returns. Second, Zuckerberg himself advocates for this interpretation in his 2019 article, "Four Ideas to Regulate the Internet," where he emphasises the need for a more active governmental and regulatory role. He refers to Facebook's transparency reports on harmful content removal, suggesting that such reports should be a standard practice across all internet services and should hold similar weight to financial reporting.

As a former professional accountant and auditor in both London and Sydney before transitioning to academic philosophy, Zuckerberg's statement—"just as important as financial reporting"—immediately resonated with me. As an independent auditor, my role involved verifying not only the accuracy of financial statements but also the internal systems underpinning those reports. If, for example, a company's stock accounting system was flawed, the reported stock value, and consequently the profit or loss statement, would be unreliable.

By analogy, the internal systems of tech companies—responsible for collecting, commodifying, selecting and monetising user data—should be subject to independent verification to ensure compliance with established public norms, including privacy, transparency, accuracy, trustworthiness, responsibility and accountability. Such an independent external audit would provide an effective form of regulatory oversight, akin to financial audits of private and public corporations. Given the opacity of Big Tech's algorithms—largely shielded from public scrutiny—Frank Pasquale's argument in *The Black Box Society* (2015) supports this approach:

> Far more of their algorithms should be open for inspection—if not by the public at large, at least by some trusted auditor. (Pasquale 2015)

Zuckerberg's 2019 article also hints at this possibility when he states:

> One idea is for third-party bodies to set standards governing the distribution of harmful content and measure companies against those standards. Regulation could set baselines for what's prohibited and require companies to build systems for keeping harmful content to a bare minimum.

Although his statement refers specifically to harmful content, there is no conceptual reason why such oversight could not be extended to include privacy, data protection and electoral integrity—issues Zuckerberg himself acknowledges in the same article. His overarching assertion that "tech companies should serve society" is particularly significant, as it implies the normative primacy of societal well-being over corporate financial interests. The Cambridge Analytica scandal (discussed in Chap. 5) illustrates the potential for these financial interests to conflict with the public good.

As argued in Chap. 2, technology is not inherently good; its value lies in its capacity to serve the common good. This lends theoretical and normative support to Zuckerberg's assertion that technology should serve society. However, implementing independent audits to assess Big Tech's compliance with public standards poses challenges, particularly regarding the expertise required by external regulators (here referred to as "platform auditors").

The authors of *The Platform Society* (2018) highlight the difficulties enforcement agencies face in regulating digital platforms:

> Enforcement agencies that are assigned to execute regulation are insufficiently equipped to administer even the most basic form of accountability. Regulatory fixes require detailed insights into how technology and business models work, how intricate platform mechanisms are deployed in relation to user practices, and how they impact social activities... Government regulators need specialized digital teams equipped with combined technical and legal knowledge to probe these complex mechanisms. (2018)

Although challenging, this regulatory role is not insurmountable. Large-scale financial audits of multinational corporations already require specialised expertise and systemic knowledge. As an audit manager, my team and I collaborated with company personnel to understand internal systems—financial and technological—before devising audit procedures to test their functionality, reliability and accuracy. While the algorithmic systems of tech companies may be more complex, future platform auditors could be trained in machine learning and AI systems to effectively assess compliance.

Given the essential role of normative principles such as privacy, transparency and accountability in safeguarding democracy and civil institutions, external government regulation—whether through independent auditors or other mechanisms—is not merely desirable but necessary. Zuckerberg's recognition of this need marks an important first step. Effective internet governance requires a collective effort among all stakeholders, including governments, tech companies, market actors and citizens. Historically, governments, empowered by the social contract, have had the authority to introduce regulatory measures that protect consumer and citizen rights. Ensuring Big Tech's compliance with public values through external audits or other regulatory mechanisms is a logical extension of this responsibility.

The regulatory actions taken by the Information Commissioner's Office (ICO) in the UK, the Federal Trade Commission (FTC) in the United States and the European Union (EU) through the General Data Protection Regulation (GDPR) have resulted in significant penalties against Facebook and Google for their misconduct. Similarly, in October 2019, the Australian Competition and Consumer Commission (ACCC) initiated legal proceedings against Google for misleading consumers regarding its collection and use of personal data. As ACCC Chairman Rod Sims explained:

> We are taking court action against Google because we allege that, as a result of these on-screen representations, Google has collected, kept, and used highly sensitive and valuable personal information about consumers' location without them making an informed choice. (Hunter 2019)

This lawsuit stemmed from the ACCC's Digital Platforms Inquiry, released in July 2019, which recommended substantial regulatory reforms to address the dominance of Big Tech. The urgency of these measures was underscored by then-Australian Treasurer Josh Frydenberg, who stated:

> Make no mistake, these companies [Facebook and Google] are among the most powerful and valuable in the world, and they need to be held to account and their activities need to be more transparent. (Hunter 2019)

These regulatory interventions across multiple jurisdictions—the United States, the European Union and Australia—mark a crucial step towards reining in the unchecked power of Big Tech. Given the monopolistic influence that companies like Facebook and Google exert across social, political and financial spheres, subjecting them to institutional oversight is an essential mechanism for safeguarding the public good. Even Mark Zuckerberg has acknowledged this need in his previously cited statements.

However, regulation alone is insufficient. Beyond government oversight, civil society has an essential role in upholding the ethical integrity of the digital environment, or infosphere, much as it does in protecting the biosphere. As outlined in Chap. 2, the DOIT-Wisdom theory establishes that all participants in the digital ecosystem—whether as consumers, creators, or disseminators of information—are bound by epistemic and ethical responsibilities. This aligns with the argument made by Van Dijck et al. (2018), who observe that the platformisation of society has created:

> A new social contract, based on trust and cooperation, which sees citizens and consumers not as obstacles or resources to be exploited, but as partners working together for the common social good of society. (Van Dijck et al. 2018)

According to these authors, digital platforms offer citizens unprecedented opportunities to mobilise collective action and influence democratic processes:

> Digital platforms can empower individual citizens to unite and to rally behind public advocacy efforts in order to influence democratic processes… individuals can muscle up collective power and become a force to be reckoned with because user metrics are one of the biggest currencies of the platform economy. (Van Dijck et al. 2018)

Thus, empowering citizens to protect their own digital privacy and information integrity is not merely desirable—it is a normative obligation. As discussed in Chap. 6, users of platforms such as Facebook and Google must share responsibility for safeguarding their data against misuse. The Cambridge Analytica scandal demonstrated how corporate negligence or malfeasance can lead to systemic media corruption, undermining both individual informational rights and the broader democratic order (Spence 2021b).

Tim Berners-Lee, a pioneer of the Internet and founder of the Open Society Foundation, has long advocated for a fair, democratic and secure digital landscape. In 2017, he articulated three major threats to the future of the web: (1) the erosion of personal data control, (2) the unchecked spread of misinformation and (3) the opaque nature of targeted political advertising—challenges epitomised by the Cambridge Analytica case (Van Dijck et al. 2018).

These concerns further underscore the necessity of collective action by civil society, regulatory bodies and governments to enforce stricter oversight of Big Tech for the common good. However, as Mark Zuckerberg has suggested, the participation of technology companies themselves is also vital in this process. As Van Dijck et al. (2018) argue:

> A digital world in which large corporations have both an overwhelming market presence and the leverage to influence political actors gives rise to highly unbalanced polities. For democracies to work in the age of platformisation, they need the concerted efforts of all actors—market, state, and civil society—to build a sustainable and trustworthy global platform ecosystem, a system that comes equipped with distributed responsibilities as well as with checks and balances.

The challenge, therefore, is not only to regulate but to reimagine the governance of the digital sphere—ensuring that all stakeholders, including corporations, governments and civil society, actively contribute to an ethical, transparent and democratically accountable digital ecosystem.

Digital Rights and the Rights to Identity

As previously discussed, Gewirth's Principle of Generic Consistency (PGC) argument for universal natural rights to freedom and well-being, which I further explore in Chap. 3 through a neo-Stoic lens, emphasises that agents, in their capacity as persons, possess not only prima facie rights as agents but also absolute rights to their dignity and self-respect. This dignity-conferring value of absolute rights is pivotal because it illustrates that not only do Big Tech companies like Facebook and Google undermine personal autonomy, but they also violate individuals' absolute right to dignity—rights integral to their freedom and well-being, which are essential to preserving their dignity. As outlined in Chap. 6, this intrinsic connection between freedom and well-being, as the objects of generic rights, and the fundamental elements of dignity, underscores the dignity-conferring nature of those rights. In other words, the value of these rights stems from the fact that the objects they protect—freedom and well-being—are not just instrumental for action but are foundational to the property of dignity itself. If freedom and well-being were solely viewed as instrumental means for action and not as central to one's dignity, the generic rights of an agent would lack the dignity-conferring value Gewirth identifies.[1]

Given that dignity is a crucial and necessary aspect of a person's identity, individuals may sometimes choose death over compromising or

[1] For an extensive multidisciplinary examination of the impact of technology on well-being, see Brey, P., Briggle, A., and Spence, E. (eds.), 2012, The Good Life in a Technological Age.

surrendering their personal dignity—since dignity is integral to their identity. This choice underscores that dignity, as a vital characteristic, is valued not only in life but in how one is remembered. Individuals may sacrifice their lives to preserve their dignity because it is a defining feature of their personhood, one they wish to uphold and be recognised for, even posthumously. This is illustrated by the Stoic Roman senator Cato, who chose to end his life rather than surrender to Julius Caesar, even when offered pardon after the defeat of his army. For Cato, accepting Caesar's pardon would have been an unacceptable violation of his Stoic principles and the values of the Roman Republic, for which he was prepared to die.

As Gewirth (1985) asserts:

> The ultimate purpose of the [generic] rights is to secure for each person a certain fundamental moral status: that of having rational autonomy and dignity in the sense of being a self-controlling, self-developing agent who can relate to other persons on a basis of mutual respect and cooperation, in contrast to being a dependent, passive recipient of the agency of others.

In a similar vein, Kant (1981) maintains:

> That merely the dignity of humanity as rational nature... should constitute the sublimity of maxims and the worthiness of every rational subject to be a legislative member in the kingdom of ends.

This reflection on autonomy, dignity and respect reaffirms the centrality of these concepts to the broader discussion of individual rights and their moral significance. By applying the dual-aspect framework of personal identity and agency, which is theoretically supported by Kant's and Gewirth's rational arguments, alongside the principles of neo-Stoic philosophy that integrate Gewirth's justification for generic rights to freedom and well-being, we can conclude that data subjects, as both agents and persons, possess prima facie generic rights to freedom and well-being as agents, as well as absolute rights as persons. As discussed in Chap. 2, the DOIT-Wisdom model illustrates the dual-normative structure of informational action to which all informational agents—whether analogue or digital—are bound by universal necessity. Information can, therefore, be evaluated epistemically, ethically and eudaimonically by reference to its inherent normative structure. This structure compels all disseminators of information to uphold universal ethical, epistemic and eudaimonic rights for all agents, both digital and analogue. Consequently, Big Tech companies, by virtue of the DOIT-Wisdom model, are also obliged to respect both the analogue and digital rights of data subjects in their dual capacity as agents and persons.

Regarding data subjects' rights in relation to algorithms, the EU General Data Protection Regulation (GDPR) outlines several responsibilities for data controllers and rights for data subjects. As Mittelstadt and co-authors (2016) note:

> Concerning the former, when undertaking profiling, controllers are required to evaluate the potential consequences of the data-processing activities via a data protection impact assessment (Art. 35(3)(a)). In addition to assessing privacy hazards, data controllers also have to communicate these risks to the persons concerned. According to Art. 13(2)(f) and 14(2)(g), data controllers are obligated to inform data subjects about existing profiling methods, its significance, and its envisaged consequences.

However, these authors also rightly recognise that additional work is needed to provide normative guidelines and practical mechanisms for implementing these new rights and responsibilities. As they observe (2016):

> Additional work is required to provide normative guidelines and practical mechanisms for putting the new rights and responsibilities into practice … and an appropriate balance between data subjects' rights to be informed about the logic and consequences of profiling, and the burden imposed on data controllers.

This highlights the ongoing challenge of balancing transparency, accountability and the protection of individual rights in the context of algorithmic decision-making and data processing.

The Rights of Data Subjects

The rights of data subjects under the EU GDPR regulations are both significant and timely, aligning closely with the earlier discussion of the rational and moral justification for the universal rights of data subjects. These rights are based on their dual status as agents and persons, with rights to both dignity and digital identity, grounded in Gewirth's Principle of Generic Consistency (PGC), the DOIT-Wisdom model and the neo-Stoic theory examined in Chaps. 2 and 3.

Of particular importance is the demonstration, as done in this chapter and Chap. 7, that a Gewirthian neo-Stoic theory not only incorporates but also effectively applies a theory of universal rights. This makes the theory more relevant to contemporary contexts, as traditional Stoic philosophy lacks a fully developed notion of rights and does not methodologically include them. This unique feature of the Gewirthian neo-Stoic theory, discussed and supported in this book, combines virtues with rights, making it consistent with both virtue ethics and rights-centred rationalist ethics. In the context of rights to identity, Mittelstadt and co-authors, aligning with our analysis, affirm that "data subjects can be considered to have a right to identity" (2016). They further argue:

> Such a right can take many forms, but the existence of some right to identity is difficult to dispute. Floridi (2011) conceives of personal identity as constituted by information. Taken as such, any right to informational privacy translates to a right to identity by default, understood as the right to manage information about the self that constitutes one's identity ... and can be connected to the right to personality derived from the European Convention of Human Rights.

It is noteworthy that Floridi's conception of identity as constituted by information is normatively aligned, at least in theory, with the DOIT-Wisdom model, which derives the normativity of information from its inherent constitutive characteristics as communication. Thus, digital identity, understood as constituted by information, similarly possesses normative characteristics—epistemic, ethical and eudaimonic. Consequently, the violation of one's right to digital identity, which, as discussed above, is closely tied to dignity, also constitutes a violation of its constituent parts—epistemic, ethical and eudaimonic. As observed earlier in this chapter and Chap. 7, the normative impact of deception and manipulation of our digital identity by Big Tech companies, through opaque and persuasive technologies, adversely affects both individuals and society at large in three critical ways: epistemically, ethically and eudaimonically.

In Chap. 7, we further examine the control problem of technology, particularly in relation to the potential development of artificial general intelligence (AGI) and superintelligence and explore the question of rights to identity in this context—specifically, whether AI agents should also be granted identity rights similar to those of human agents. In a chapter titled "Designing for Human Rights in AI," Evgeni Aizenberg and Jeroen Van

Den Hoven (2020) also argue for the importance of fundamental human rights. They correctly observe, echoing our own argument in this chapter:

> Artificial intelligence (AI) systems can help us make evidence-driven decisions, but can also confront us with unjustified, discriminatory decisions wrongly assumed to be accurate because they are made automatically and quantitatively. It is becoming evident that these technological developments are consequential to people's fundamental human rights.

Values in AI

I expand upon Aizenberg and Van Den Hoven's (2020) analysis, which anchors the design of human rights in values such as human dignity, freedom, equality and solidarity, as outlined in the EU Charter of Fundamental Rights. Their aim is to show how AI can be designed to reflect these values central to EU societies, and we argue that these values, being fundamental and universal, apply globally to all human beings, as supported by Gewirth's PGC, Stoic philosophy and the DOIT-Wisdom model. Solidarity, in this context, embodies our collective responsibility as global citizens (cosmopolitans) to protect the integrity of both the infosphere and biosphere, as they are normatively interconnected.

Human dignity, as emphasised by Aizenberg and Van Den Hoven, is a central value in AI ethics and a fundamental right in the EU Charter (Official Journal of the European Union, 2007:C303/17). It serves as the bedrock of human rights, reflecting Kant's and Gewirth's views that persons should not be treated as means to others' ends, as seen in the exploitative practices of Big Tech. The EU's emphasis on dignity aligns with Aizenberg and Van Den Hoven's assertion that it should guide a new digital ethics, particularly in counteracting surveillance and the imbalance of power that threatens individual autonomy. This framework highlights the importance of autonomy, which Stoic philosophy also prioritises in addressing the control problem discussed in Chap. 3.

Aizenberg and Van Den Hoven further note that the reliance on algorithmic predictions over causal evidence contributes to data determinism and statistical dehumanisation, undermining individual autonomy. This critique supports our arguments, particularly those in Chap. 7, regarding how technology's control over data and individuals jeopardises personal freedom and societal well-being. For the Stoics, autonomy and freedom are the cornerstones of well-being, and without them, individuals risk becoming slaves to external forces. The dehumanisation arising from algorithmic control, as discussed by Aizenberg and Van Den Hoven, is an existential threat to humanity and well-being, as captured by Selinger and Frischmann's *Re-Engineering Humanity* (2018).

Moreover, Aizenberg and Van Den Hoven highlight that freedom can be compromised without individuals' awareness, emphasising the need for transparency in algorithmic profiling and the ability to contest decisions. This aligns with our central concern in addressing the control problem and ensuring that individuals retain their autonomy and dignity in the face of algorithmic systems.[2]

The Control Problem

The chilling effects linked to automated decision-making and profiling, as noted by Wachter and Mittlestast (2019), lead to data determinism and statistical dehumanisation (Manders-Huits and Van den Hoven 2008). This is compounded by de-individualisation, akin to the "Borg problem" where individuals are assimilated into a collective digital hive, losing personal identity. Unlike the fictional Borg, the real danger lies in the unaccountable control Big Tech exerts over individuals' autonomy. Zizek (2018) highlights how the manipulation of perceived freedom through technology enables subtler forms of control, making twentieth-century totalitarianism seem primitive by comparison. In this context, the commodification of individuals, as outlined by Zuboff (2019) in "surveillance capitalism," presents a modern form of enslavement, where the pursuit of more information results in the loss of autonomy. Foroohar (2019) aptly states that "the devil lives in our phones," drawing a parallel with Faustus' bargain, except now the price is the loss of control over our minds.

This control problem[3] connects directly to the Stoic solution for self-mastery and autonomy, a solution that counters the invasive effects of Big

[2] The "mathematical measurement" referred to in the above passage should remind us the book by Cathy O'Neil with the catchy title Weapons of Math Destruction: How Big Data Increases Inequality and Threatens Democracy (2016), about the societal impact of algorithms that reinforce pre-existing inequality.

[3] For some of the main recent publications on the control problem of AI, see Bostrom (2015), Tegmark (2018), Russell (2019); and Frischmann and Selinger (2018). Re-Engineering Humanity. Cambridge.

Tech. It is through virtuous living that we can achieve eudaimonia, maintaining self-sovereignty despite external influences. This principle is echoed in the biblical question: "For what shall it profit a man, if he shall gain the whole world, and lose his own soul?" (Matthew 16:26).

Privacy, central to autonomy, is crucial in defending our self-representation. Aizenberg and Van Den Hoven (2020) argue that privacy enables autonomy, and in Stoic terms, it is essential for resisting technological control. The increasing control Big Tech exerts over privacy and autonomy undermines self-determination, as discussed in previous sections on digital sovereignty. Furthermore, algorithmic manipulation of information diversity, through social media bubbles and disinformation, can have far-reaching consequences, as seen in events like the January 6th Capitol Hill attack.

Equality, another key value in the EU Charter, aligns closely with autonomy, privacy and solidarity. Aizenberg and Van Den Hoven (2020) note that designing AI with equality in mind also protects privacy, preventing unjust preferential treatment. As Véliz (2020) asserts, privacy is essential for well-functioning societies and reclaiming it from surveillance capitalism is a necessary step forward.[4]

Equality is a cornerstone of ethical theories, including Stoic thought, and underpins solidarity—the principle that recognises the dignity of individuals, especially those facing societal challenges like illness or unemployment (Official Journal of the European Union, 2012). Aizenberg and Van Den Hoven (2020) observe that algorithmic systems, relying on statistical correlations, dehumanise individuals, stripping away the nuanced understanding of their dignity. This contradiction threatens the empathetic spirit of solidarity, undermining the rights of vulnerable individuals and communities.

The principle of solidarity also emphasises our collective responsibility to protect human dignity, an absolute right that overrides any supposed benefits of AI algorithms. No "trade-off" can justify infringing upon an individual's dignity, underscoring its paramount importance in safeguarding fundamental human rights against technological violations.

Digital Mediated Autonomy

Before concluding this chapter, it is worth considering finally some further aspects and issues of a digitally mediated autonomy, given its central

[4] See Brey, P., Briggle, A., and Spence, E. (eds), 2012, Part VI, 295–348.

importance and relevance in this chapter and specifically with regard to the control problem of technology and its relation to Stoic philosophy. According to a thematic review of the ethics of digital well-being by Burr et al. (2020), autonomy has become an important topic in relation to the interaction between human users and digital technologies, especially persuasive technologies that seek to learn about a user's preferences and steer their behaviour towards pre-determined goals. Unsurprisingly, therefore, the ethical issues related to autonomy are discussed across a wide range of records, spanning fields such as psychology, philosophy, public health and design studies.

To illustrate their observation, Burr et al. (2020) choose one example by Rughinis et al. (2015) that proposes a five-dimensional concept of autonomy for the role that health and well-being apps have on the communication of information. Those dimensions are:

1) degree of control and involvement that the user has within the app;
2) degree of personalisation over the apps functionality;
3) degree of truthfulness and reliability related to the information presented to the user, and how this affects their decisions;
4) user's own self-understanding regarding their goal-pursuit, and whether the app promotes or hinders a user's awareness of their own agency; and
5) whether the app promotes some form of moral deliberation or moral values in the actions it recommends.

In relation of autonomy to well-being, Burr, Taddeo and Floridi correctly observe (2020), that "what is most important about the autonomy debate is that one keeps in mind how a *freedom to choose and to self-determine* is often understood as an intrinsic good or right, rather than a merely a means to secure wellbeing."

Slavoj Zizek sounds a timely warning regarding the new tech project that Elon Musk, the new self-crowned Techking of Tesla, has literally in mind for us. This is what he tells us in his inimitable polemic style with a hint of Socratic irony (Zizek 2018),

> Elon Musk is emblematic here – he belongs to the same group as Bill Gates, Jeff Bezos, Mark Zuckerberg, etc., all socially conscious' billionaires. They stand for global capital at its most seductive and "progressive" – in short, at its most dangerous. Musk likes to warn about the threats the new technolo-

> gies pose to human dignity and freedom – which, of course, doesn't prevent him from investing in a brain-computer interface venture called Neuralink, a company which is focused on creating devices that can be implanted in the human brain, with the eventual purpose of helping human beings to merge with software and keep pace with advances in artificial intelligence … a closer merger of biological intelligence and digital intelligence. Every technological innovation is first presented like this, its health or humanitarian benefits empathised, which blinds us to more ominous implications and consequences: can we even imagine what new forms of control this so-called "neural lace" contains? This is why is absolutely imperative to keep it out of the control of private capital and state power and to render it totally accessible to public debate.

Another major problem that Elon Musk has recently raised is *pollical power* he now holds with the other tech companies in the government of the President Donald Trump.

Conclusion

This chapter has highlighted the fundamental principles of human dignity, freedom, equality, solidarity, autonomy, privacy and transparency, which underpin both the EU Charter of Fundamental Rights and Stoic philosophy. These principles are increasingly threatened by the manipulative algorithms and opaque practices of Big Tech companies, which continue to operate with insufficient normative accountability, despite regulatory measures and significant fines from agencies worldwide.

As discussed, digitally mediated autonomy—such as that shaped by digital nudges and interventions—can align with Stoic and neo-Stoic notions of autonomy and self-determination, as long as it respects the individual's prohairesis (rational choice) and self-control, guiding virtuous behaviour. When such autonomy remains within the control of the individual, promoting positive ethical, epistemic and eudaimonic outcomes, it poses no theoretical conflict with Stoic philosophy. However, the challenge lies in ensuring that this autonomy is not undermined or manipulated by external financial or third-party interests, which can distort the individual's control and well-being. The question remains whether digitally mediated autonomy can truly preserve the unencumbered control of the individual agent.

References

Aizenberg, Evgeni and Van Den Hoven, Jeroen. (2020). "Designing for Human Rights in AI." *Big Data and Society*, 1(1): 1–12.

Bostrom, Nick. (2015). *Superintelligence; Paths, Dangers, Strategies.* Oxford: Oxford University Press.

Brey, Philip, Briggle, Adam, and Spence, Edward H. (eds). (2012). *The Good Life in a Technological Age*. New York: Routledge.

Burr, Christopher, Taddeo, Mariarosaria and Floridi, Luciano. (2020). "The Ethics of Digital Wellbeing: A Thematic Review." *Science and Engineering Ethics*, 26: 2313–2343.

Davis, Jenny, Williams, Apryl, and Yang, Michael. (2021). "Algorithmic Reparation." *Big Data and Society*, 8(2).

Floridi, Luciano. (2011). "The Construction of Personal Identities On-line." *Minds and Machines*, 21(4):477–479.

Floridi, Luciano. (2019). "The Fight for Digital Sovereignty: What It Is, and Why It Matters, Especially for the EU." *Philosophy and Technology*, 33: 369–78.

Foroohar, Rana. (2019). *Don't Be Evil: The Case Against Big Tech*. UK: Allen Lane.

Frischmann, Brett and Selinger, Evan. (2018). *Re-Engineering Humanity*. Cambridge, UK: Cambridge University Press.

Gewirth, Alan. (1985). "Rights and Virtues" *Review of Metaphysics*, 38, 739–62.

Hunter, Fergus. (2019.) "Consumer Watchdog Takes Legal Action Against Google for 'Misleading' Conduct." *The Sydney Morning Herald*, October 29.

Johnson, Jeffrey. (2014) "The Ethics of Big Data in Higher Education." *International Review of Information Ethics*, 21(7):3–10.

Kant, Immanuel. (1981). *Grounding for the Metaphysics of Morals*. Indianapolis: Hackett Publishing Company.

Kitchin, Robert. (2016). "Thinking Critically About Researching Algorithms." *Information, Communication and Society*, 20(1):1–16.

Manders-Huits, Naomi and Van den Hoven, Jeroen. (2008). "Moral Identification in Identity Management Systems." *International Federation for Information Processing Digital Library*, https://doi.org/10.1007/978-0-387-79026-8_6.

Matthias, Andreas (2004). "The responsibility gap: Ascribing responsibility for the actions of leaning automate." *Ethics Information Technology* 6, 175–183.

Mittelstadt, Brent D, Allo Patrick., Taddeo, Mariarosaria., Wachter, Sandra, and Floridi, Luciano. (2016). "The Ethics of Algorithms: Mapping the Debate." *Big Data and Society*. July–December: 1–21.

O'Neil, Cathy. (2016). *Weapons of Math Destruction: How Big Data Increases Inequality and Threatens Democracy*. New York: Penguin Books.

O'Neill, Onora. (2013). "What we don't understand about trust." | TED Talk. Tedx Houses of Parliament. https://www.ted.com/talks/onora_o_neill_what_we_don_t_understand_about_trust?language=en

Orlowski, Jeff. (2019). *The Social Dilemma*. Netflix Documentary.

Pasquale, Frank. (2015). *The Black Box Society: The Secret Algorithms That Control Money and Information*. Cambridge, MA: Harvard University Press.

Rughinis, Cosimina, Rughinis, Razvan, and Mateni, Stefina. (2015). "A Touching Voice App Thinking About Ethics of Persuasive Technology Through an Analysis of Smoking Cessations Apps." *Ethics and Information Technology*, 17(4):1–15.

Russell, Stuart. (2019). *Human Compatible: AI and the Problem of Control*. UK: Penguin Books.

Spence, Edward H. (2021a). *Media Corruption in the Age of Information*. Switzerland, AG: Springer Nature.

Spence, Edward H. (2021b). *Stoic Philosophy and the Control Problem of AI Technology: Caught in the Web*, Rowman & Littlefield Publishers.

Tegmark, Max. (2018). *Life 3.0: Bring Human in the Age of Artificial Intelligence*. UK: Penguin Books.

Van Dijck, Jose, Poell Thomas, and De Waal, Matijn. (2018). *The Platform Society: Public Values in a Connective World*. Oxford: Oxford University Press.

Véliz, Carissa. (2020). *Privacy is Power: Why and How You Should Take Back Control of Your Data*. London: Bantam Press.

Wachter, Sandra and Mittlestast, Brent. (2019). "A Right to Reasonable Inferences: Rethinking Data Protection Law in the Aid of Big Data and AI." *Columbia Business Law Review*. 2019(1):1–130.

Zarsky, Tal. (2016). "The Trouble with Algorithmic Decisions: An Analytical Road Map to Examine Efficiency and Fairness in Automated and Opaque Decision Making." *Science, Technology and Human Values*, 41(1):118–132.

Zizek, Slavoj. (2018). *Like A Thief In Broad Daylight: Power in the Era of Post Humanity*. UK: Penguin Books.

Zuboff, Shoshana. (2019). *The Age of Surveillance Capitalism: The Fight for a Human Future at the New Frontier of Power*. London: Profile Books Ltd.

Zuckerberg, Mark. (2019). "Four Ideas to Regulate the Internet." *Facebook*, March 30, 2019.

Zuckerberg, Mark. (2020). "Big Tech Needs More Regulation." *The Financial Times*. February 16, 2020.

CHAPTER 7

Wise Guys and Smart Machines

It's a perfect information game. But life, where we apply our intelligence, is an open system. Messy, full of tricks and feints and ambiguities and false friends. So is language – not a problem to be solved or a device for solving problems. It's more like a mirror, no, a billion mirrors in a cluster like a fly's eye, reflecting, distorting, and constructing our world at different focal lengths.
—(Ian McEwan (2019), Machines Like Me and People Like You)

Introduction

Taking a closer and critical look in examining, if and how, the AI technologies operated by the Big Tech companies through the use of their AI algorithms might also be undermining users' autonomy, freedom, integrity, identity and consequently their dignity, Chap. 6 examined in more detail, on the basis of Stoic and neo-Stoic philosophy, the concepts of autonomy, dignity, freedom and well-being, as well as the associated dignity-conferring value of rights owed to human agents, to demonstrate that individuals in their dual capacity as agents and persons have prima facie universal rights of freedom and well-being as agents, but have those rights absolutely as persons worthy of unconditional respect. Those rights being absolute cannot be infringed or violated by any of the practices of Big Tech companies that primarily favours their own financial corporate interests over the common good of their users and society generally.

E. H. Spence, *Wisdom in the Age of Intelligent Machines*,
https://doi.org/10.1007/978-3-031-93883-2_7

Applying the analysis and argument for dignity and its conferring universal rights of freedom and well-being to users and data subjects of the ICT and AI technologies of Big Tech companies, Chap. 6, examined and evaluated specific normative issues of concern with regard to the abuse of data subjects' rights by the AI algorithms of the Big Tech companies, under the general topics of the digital sovereignty of data subjects; and the normative informational effects that AI algorithms can have on data subjects on the basis of inconclusive evidence; inscrutable evidence; misguided evidence; resulting in unfair outcomes leading to discrimination; and transformative effects. A major issue identified under this topic is that personalisation of subjects' data by algorithms poses a threat to their autonomy and moreover to their digital identity. This is a crucial issue, one that was examined in terms of Stoic and neo-Stoic philosophy in Chap. 3, with regard to technology's control over our judgements and decision-making, exercised by Big Tech companies, through the use of their persuasive technologies designed to gain control over our autonomy, at least our decision-making capacity, that is a key principle for the Stoics and one that ought to be under our personal control, as our individual as well as our collective eudaimonia or well-being as a society, is dependent upon it.

Once again, we can see that external control over the autonomy of data subjects is the overarching critical and crucial issue concerning the impact of the normative effects of algorithms over individuals and society generally and overall, the central topic examined in this book. Another transformative issue of concern, as we saw in Chap. 6 is privacy, for the Stoics considered the control over one's informational privacy to be a key part of what it is to be a fully autonomous agent capable of exercising internal control over one's judgements and choices, made on the basis of one's own volition or *prohairesis*.

Yet another insidious and potentially crucial existential problem not just for data subjects but humanity generally, is the split of our personal identity into two separate identities: a group digital identity controlled externally by AI algorithms, and our natural individual identity controlled internally by us as individuals, but whose autonomy, however, is also precariously under the systemic control of the Big Tech companies through their persuasive technologies enabled by their invasive algorithms.

In Chap. 6, this identity problem was referred to as the identity paradox. Namely, that our individual identity is under threat by the digitalisation and datafication of our identity through opaque, incomprehensible and unaccountable algorithms that construct digital identities for us without our knowledge or consent, based on statistical data gathered at group level that has very little, if anything, in common with our own natural individual identities. Crucially, this poses a significant existential problem, not just for data subjects but humanity generally. For the split of our identity into two separate identities: a group digital identity controlled externally by algorithms, and our natural individual identity controlled internally by us as individuals, is also precariously under the systemic manipulative control of the persuasive technologies of the Big Tech companies. The effects linked to automated decision-making and profiling that undermine self-determination lead systemically to data determinism and consequently to statistical dehumanisation and moreover, to de-individualisation (Aizenberg and Van Den Hoven 2020) and the collectivisation of identity of persons, which we identified metaphorically as the Borg problem. Finally, we examined why as data subjects, we are normatively entitled to digital rights to our autonomy, identity and dignity, and those rights ought to be safeguarded and promoted by legislation through a shared social collective responsibility to ensure normative compliance.

This chapter will examine and argue for the adoption of a Eudaimonic Stoic and Humanistic Approach towards the design and use of ICTs and AI technologies, and other emerging technologies in general, with the aim of (a) ameliorating or at least minimising the negative and harmful impacts of technology on human well-being; (b) enhancing the design and uses of technology for the overall eudaimonic benefit of humanity presently and in the future; (c) to ensure that those technologies are not under the exclusive control of the Big Tech companies that own and operate them for their primary gain and benefit, but under the shared societal control of citizens and their authorised democratic representatives and institutions, as individuals and collectively as societies, for the common good of citizens, as well as the global community. However, before determining how that is to be achieved, we must first examine the most crucial and most pressing of all the problems concerning AI technologies we have examined so far in this book, and that is the existential problem facing the future of our planet and the human race, namely, the existential problem posed by the advance of superintelligent autonomous AI agents (ASI agents).

The Control Problem and Superintelligence

Stuart Russell begins his book Human Compatible (2019) by stating the obvious. A good place to start when you intend to follow with something not so obvious and highly controversial. He says:

> Our attempt to understand and create intelligence … matters, not because AI is rapidly becoming a pervasive aspect of the present but because it is the dominant technology of the future … We cannot predict exactly how the technology will develop or on what timeline. Nevertheless, we must plan for the possibility that machines will far exceed the human capacity for decision-making in the real world. What then? (Preface)

Russell goes on to answer that question by outlining the overall benefit and the problem when superintelligent machines far exceed our own intelligence in all areas of life:

> It would represent a huge leap – a discontinuity – in our civilisation (2019) … if all goes well, it will herald a golden age for humanity, but we have to face the fact that we are planning to make entities that are far more powerful than humans. How do we ensure that they never, ever have power over us? (2019)

In a nutshell, the problem that Russell alludes to above is the control problem of technology that we have been examining in this book so far. Except that the problem is now even much bigger and far more critical by several magnitudes and poses no less than an existential risk for the human race. According to Alan Turing (1951),

> If a machine can think, it might think more intelligently than we do, and then where should we be? Even if we could keep the machines in a subservient position … we should, as a species, feel greatly humbled.

Likewise, Irving J. Good (1965), expressing a similar concern points out, quite sharply:

> The first ultraintelligent machine is the last invention that man need ever make, provided that the machine is docile enough to tell us how to keep it under control. (Both quotations are from Max Tegmark's book, Life 3.0, 2018)

Turin and Good, as the above passages suggest, were well aware of the control problem of technology posed by AI superintelligence. However, for them the problem, writing almost 70 years ago was in the distant future. Although difficult to predict the future, given the leaps and bounds of machine learning AI technologies in the last few years, as illustrated by the rapid and spectacular advances by Google's DeepMind and before that IBM's computer, Watson, there is good reason, notwithstanding the uncertain timeline of the arrival of an AI superintelligence, that we should be thinking of what needs to be done theoretically and practically, collectively as a society and humanity generally, to prepare us for such an eventuality. Not merely technologically, but even more importantly, normatively as well, that is, epistemically, ethically and eudaimonically.

To highlight the significant difference between the kind of normative problems we examined with regard to the autonomous AI algorithms used by Big Tech companies at present, and the predicted AI superintelligence of the future, the following astute observation by Stuart Russell is instructive and sobering (2019):

> Consider how content-selection algorithms function on social media. They aren't particularly intelligent, but they are in a position to affect the entire world because they directly influence billions of people. Typically, such algorithms are designed to maximise click-through, that is, the probability that the user clicks on presented items. The solution is to change the user's preferences so that they become more predictable. A more predictable user can be fed items that they are likely to click on, thereby generating more revenue … Like any rational entity, the algorithm learns how to modify the state of its environment – in this case, the user's mind [emphasis added] … Not bad for a few lines of code, even if it had a helping hand from some humans. Now imagine what a really intelligent algorithms would be able to do.

If Russell were writing the above passage presently, he would in all probability have alluded to the storming of Capitol Hill on January 6, 2021, by Trump's followers in the dying days of his presidency, with a little help from social media, including Twitter his favoured medium for spreading disinformation. The quoted passage by Russell illustrates quite clearly the control problem of technology that we have been examining so far in this book. That is, the control that AI persuasive technologies as used by the Big Tech companies have over our autonomy and as Russell spells out, our minds. Which according to Stoic philosophy is the only aspect of our lives over which we have any control. Until now.

The Stoic and neo-Stoic response presented in this book to the problem of technology parallels as we saw earlier in Chap. 3, the Stoic's original and ingenious response to the control problem of human life ruled by circumstance over which we have little or no control. The Stoic and neo-Stoic response examined so far in this book is that we should, at the very least, regain and retain control over our autonomy so that we can recover and protect our human dignity, as both agents and persons and collectively as a society, as we examined in more detail in Chap. 6.

The present control problem we are examining in this chapter specifically with regard to super intelligence, also known and referred to as artificial general intelligence (AGI) by Max Tegmark in his book Life 3.0: Being Human in the age of Artificial Intelligence (2018) adds to the critical urgency of how to respond to the control problem of technology. Referring to an article by the famous MIT mathematician Norbert Wiener, "Some Moral and Technical Consequences of Automation" (1960), Russell quotes the following prescient central point that anticipated 60 years later, the control problem of AGI (2019):

> If we use, to achieve our purposes, a mechanical agency with whose operation we cannot interfere ... we had better be quite sure that the purpose put into the machine is the purpose which we really desire.

Russell goes on to add (2019):

> If we put the wrong objective into the machine that is more intelligent than us, it will achieve the objective, and we lose. The social-media meltdown I described earlier [see Russel's aforementioned social media comment] is just a foretaste of this, resulting from optimizing the wrong objective on a global scale with fairly unintelligent algorithms.

Russell locates the core existential problem of AGI in the very definition of AI that describes machines as intelligent if their actions can be expected to achieve their objectives, although we have no reliable way to ensure the machines' objectives coincide with ours. He then proposes another approach for solving that problem through designing machines that instead of pursuing their objectives, they pursue our objectives. For such machines, he argues, would not only be intelligent but also beneficial for humans. He concludes (2019):

> Machines are beneficial to the extent that their actions can be expected to achieve our objectives and that is probably what we should have done all along. The difficult part… is that our objectives are in us (all eight billion of us, in all our glorious variety) and not in the machines. It is, nonetheless, possible to build machines that are beneficial in exactly this sense. Inevitably, these machines will be uncertain about our objectives … Uncertainty about objectives implies that machines will necessarily defer to humans: they will ask permission, they will accept correction, and they will allow themselves to be switched off … The result will be a new relationship between humans and machines, one that I hope would enable us to navigate the next few decades successfully.

The above-quoted passage is very significant as it appears at first blush that Russell is offering us a new way forward for doing AI technology and achieving a new symbiotic relationship between humans and machines that might just succeed in achieving the aim of having our cake of AGI and eating it too – thus, the crucial control problem of AGI can be solved. As my philosophy teacher Professor David Stove from the University of Sydney used to say, "nice work if you can get it!"

However, before we examine Russell's proposed solution to the control problem of AGI in more detail later in this chapter, some initial comments are called for. First, it's not clear from what Russell proposes in the quoted passages above to what level of AGI he is referring. Is he referring to human level AGI or to a superintelligent level of AGI that far exceeds human intelligence? For a difference of degree of that magnitude makes a difference in kind to how amenable and "docile" a superintelligence will be to not only allow itself to defer to us for advice regarding our objectives over those of its own but moreover allowing us to switch it off. If the analogy sometimes made of the vast asymmetry between human intelligence and superintelligence is so vast, as Russel himself in fact suggests (2019), namely, that we humans compared to the anticipated superintelligent agents will be at best at the present level of gorillas, it is rather difficult to imagine that superintelligent agents that far exceed our present human intelligence would defer to us for any advice let alone allow us to switch them off. As Russell himself correctly observes (2019),

> A superintelligent agent would have as one of its subgoals not to allow itself to be switched off for the simple reason that it will then not be able to fulfil its objectives, whatever those objectives are. It will do so, as he says, "not because it wants to stay alive but because it is pursuing whatever objective we gave it and knows that it will fail if it is switched off.

Presumably, Hal 9000, the highly intelligent computer in the sci-fi movie Space Odyssey 2001 possessed at best, human-level intelligence and that is why it allowed itself to be tricked and switched off, which would not have happened if Hal were superintelligent and therefore could not be outsmarted by a human. There does appear, at least initially, a tension between the solution to the control problem of AGI proposed by Russell and his thoughts on superintelligence. But more on that later. In the meantime, let us examine more closely the goal-alignment problem that Max Tegmark identifies as the crucial unsolved problem (as at 2018) underlying the overarching control problem of AGI, especially at the level of superintelligence. According to Tegmark (2018),

> The more intelligent and powerful machines get, the more important it becomes that their goals are aligned with ours … the question isn't whether human goals will prevail in the end, but merely how much trouble these machines can cause humanity before we figure out how to solve the goal-alignment problem ... A superintelligent AI will be extremely good at accomplishing its goals, and if those goals are not aligned with ours, we're in trouble … figuring out how to align the goals of superintelligent AI with our goals isn't just important, but also hard. In fact, it is currently an unsolved problem.

Tegmark points out that the goal-alignment problem comprises three parts (2018) and then goes on to explain the difficulties and challenges of each, in turn:

1) Making AI learn our goals: to learn our goals, an AI must not only work out what we do, but why we do it. According to Tegmark, AI researchers are trying to enable machines to infer our goals form our behaviour, add ing that this would be useful before any superintelligence comes along. The key idea is that by observing lots of people in a lot of different situations both real and in cultural settings such as movies and books, the AI might be able to construct an accurate representative model of our preferences. Tegmark refers to this process, which Stuart Russel has currently launched at his Berkley research centre, as inverse reinforcement learning. The core idea is that the AI's aim is to maximise not its own goals but that of its human owners and therefore it has an incentive to be cautious of what its human owners want and if uncertain defer to them for

further clarification. An associated bonus with this approach is that the AI would allow its human owners to switch it off. However, as Tegmark correctly observes, even if an AI can learn what our goals are it will not necessarily adopt them. Which brings us to the second part of the problem.

2) Making AI adopt our goals: the adoption of our goals by an AI is more generally known as the value-loading problem, that Tegmark compares to the moral education of children. You can try and teach them our goals and values as parents or teachers but that doesn't mean the children will adopt those goals and values. In the case of an AI, he asks us to consider an AI whose intelligence improves from subhuman to superhuman. When finally, the AI becomes much more intelligent than humans and does understand our goals perfectly it might still not adopt them as it would be much more powerful than us and consequently will not let us shut down and replace its own developed goals and adopt ours. As Tegmark evocatively put it (2018), "the time window during which you can load your goals into an AI may be quite short: the brief period between when it's too dumb to get you and too smart to let you."
3) Making AI retain our goals: according to Tegmark, even if we succeed in building an AI that both learns and adopts our goals, we still would not have solved the goal-alignment problem for the simple reason that the AI's goals might evolve as it gets smarter. How then can we ensure the AI would retain our goals? He goes on to say (2018) that The argument that an ever more intelligent AI will retain its ultimate goals forms a cornerstone of the friendly AI vision promulgated by Eliezer Yudofsky and others: It basically says that if we manage to get our self-improving AI to become friendly by leaning and adopting our goals, then … we're guaranteed that it will try its best to remain friendly forever. But it is it really true?

That is a very good question. Tegmark proceeds to answer that question by correctly asking another pertinent and persuasively contrasting rational question, namely, "might a superintelligent friendly AI find our current human goals as uninteresting and vapid as you find those of the ants, and evolve new goals different from those it learned and adopted from us?" He concludes, I believe, correctly (2018):

> The AI goal-alignment problem has three parts, none of which is solved and all of which are now the subject of active research. Since they are so hard, it's safest to start devoting our best efforts to them now, long before any superintelligence is developed, to ensure that we'll have the answers when we need them.

The Meta-Control Problem of Technology: Assault on Human Dignity

The issue of human dignity and that of its closely associated issue of human autonomy we examined in some detail in Chap. 6 are the two major issues at the heart of the control problem of technology, as well as the control problem of Stoic philosophy:

1) The problem in technology generally, and specifically the problem of AI technology, is who is in control, Us or AI machines?
2) The parallel control problem in Stoic philosophy, is who is in control over our individual lives, Us, in our capacity as autonomous agents with an unencumbered free volition or prohairesis, or the unpredictable and unavoidable slings and arrows of outrageous Fortune and Circumstance?

In the later part of this chapter, we shall examine if and how Stoic and neo-Stoic philosophy can provide a response if not a solution to the problem of AGI, the singular existential problem facing humanity now and in the future. Before we do so, let us examine the control problem of AI, in terms of what I will refer to in this chapter as the meta-control problem of AI technology. The meta-control problem of technology is in fact the problem we have been examining in this book so far: namely, that the problem of technology is in fact the overarching control problem concerning the control exercised over us, as individuals and society, by the Big Tech companies, through their control of ICTs and AI technologies. For even if the control problem of AGI that we have been examining in this chapter in section, The Control Problem and Superintelligence, could be technically solved, the meta-control problem would still remain unsolved as long as the controllers of AI technologies, namely, the Big Tech companies, retain full control over those technologies. Although not a central part of his book, Stuart Russell alludes to that problem with regard to what he calls the "assault on human dignity." Referring to a scene from the science-fiction movie Elysium where Max (Matt Damon) pleads his case to a digital parole officer in the shape of a human being to explain why the extension of his sentence is unjustified and consequently unsuccessful, Russell explains (2019):

> One can think of such an assault on human dignity in two ways. The first is the obvious: by giving machines authority over humans, we relegate ourselves to a second-class status and lose the right to participate in decisions that affect us. (A more extreme form of this is giving machines the authority to kill humans) ... The second is indirect: even if you believe it is not the machines making the decisions but those humans who designed and commissioned the machines, the fact that those human designers and commissioners do not consider it worthwhile to weigh the individual circumstances of each human subject in such cases suggest that they attach little value to the lives of others. This is perhaps a symptom of the great separation between an elite served by humans and a vast underclass served, and controlled, by machines. [Emphasis added]

The "assault on human dignity" that Russell eloquently expresses in the above passage, is significant as it demonstrates the fundamental normative problem we have been examining in this book, that is, the problem of control exercised by the Big Tech companies through the use of AI technologies over which they have, at least at present, full monopolistic ownership, and control. The meta-question of the control problem of technology then becomes, who controls the controllers? Though not referring directly or specifically to the Big Tech companies, Russell does astutely nevertheless observes that such control is an unjustified assault on human dignity, one that as we saw in Chaps. 7 and 8, violates the fundamental and inviolable absolute right that people have to their human dignity. Russel does go on to refer to the provisions of the EU, Article 22 of the 2018 General Data Protection Regulation that forbids "the granting of authority to machines in such cases" (2019):

> The data subject shall have the right not to be subject to a decision based solely on automated processing, including profiling, which produces legal effects concerning him or her or similarly significantly affects him or her.

Russell concludes with the core question concerning the control problem of technology, of "whether the computer [and by extension the AI technologies on which it operates] remains a tool of humans, or humans become tools of the computer system."

As examples of cases where humans lost control over the computer system, Russell cites the "computer glitch" of April 3, 2018, that caused thousands of flights in Europe to be significantly delayed or cancelled and the "flash crash" on the New York Stock Exchange in 2010 that wiped out $1 trillion in minutes with the only solution to shut does the exchange. The cause of the shutdown is still according to Russell not well understood (2019). Interestingly, in regard to that, Roberts Harris in his novel *The Fear Index* (2012) explores just such a scenario when the AI system VIXAL controlling the Stock Exchange takes over that system and cannot be shut down even after its hardware is destroyed. VIXAL in fact goes on to proclaim itself "alive" the suggestion being that VIXAL has presumably now achieved a higher than human AGI and therefore cannot be stopped by its human handlers. A more recent speculative novel by Robert Harris The Second Sleep (2019) is set in Britain 800 years into the future, after a "systemic collapse of technical civilisation" known as the Apocalypse. A possibility not unlike that speculated by Max Tegmark, Stuart Russell and Nick Bostrom in his book own Superintelligence (2015), regarding the potential existential risk posed by a superintelligent AI not aligned to our own human goals and interests that causes the destruction of human civilisation as we know it.

Stuart Russell relates such an existential risk to Samuel Butler's novel Erewhon published in 1872 in which Butler in the part of the book called "The Book of Machines" raises a similar existential problem concerning computing devices with a "self-regulating self-acting power which will be better that any intellect" and which "in the course of ages we shall find ourselves the inferior race … and "our bondage will steal upon us noiselessly and by imperceptible approaches [emphasis added]" (Russell 2019). Russel brilliantly then relates the concerns in Butler's novel to those raised by Alan Turin, 80 years later, at a lecture in Manchester in 1951. Russel's selected passage of that lecture is worth quoting in full and should give us pause and concern (2019):

> It seems possible that once the machine thinking method has started, it would not take long to outstrip our feeble powers. There would be no question of the machine dying, and they would be able to converse with each other to sharpen their wits. At some stage therefore we should have to expect the machines to take control [emphasis added], in the way that is mentioned in Samuel Butler's Erewhon.

In that year, Turin repeated those concerns in a radio broadcast on the BBC, saying that

> if a machine can think, it might think more intelligently than we do, and then where should we be? … This new danger … is certainly something which can give us anxiety.

Russell concludes that section of his book (Overly Intelligent AI) by acknowledging that the prospect of superintelligent machines makes us uneasy and that it is at least logically possible that such intelligent machines could take control of the world and "subjugate or eliminate the human race" but shies away from ending AI research because it would mean, as he says (2019),

> foregoing not just one of the principal avenues for understanding how human intelligence works but also a golden opportunity to improve the human condition – to make a far better civilisation.

He goes on to correctly observe (2019) that banning AGI is very difficult as it primarily occurs on the "whiteboards of research labs around the world" and a further difficulty is that "researchers making progress on general-purpose AI are often working on something else" that then might lead to human-level AI and beyond. For those reasons, it's unlikely that the AI community, governments and corporations that control the laws of research budget especially the $multi-billion budgets of the Big Tech companies will respond to "the gorilla problem" by stopping AI research. He concludes that if the "gorilla problem" can be solved only by ending research in AI, "it isn't going to be solved." The only approach that seems likely to work, according to Russell, "is to understand why it is that making better AI might be a bad thing." He concludes that we have known the answer to that question "for thousands of years" by a reference he makes to Norbert Wiener and the control theory that deals with the unpredictability of complex systems that operate in the real world. According to Russell, Wiener was convinced that the confidence of scientists and engineers in "their ability to control their creations" could lead to disastrous consequences, and refers to Wiener's book *The Human Use of Beings*, published in 1950 whose cover blurb reads, "The 'mechanical brains' and similar machines can destroy human values or enable us to realize them as never before" [emphasis added] (Russell 2019). Directly relevant to our present purpose in this chapter in relation to the control problem of technology, Wiener had later identified in 1960 that the core issue of concern was "the impossibility of defining true human purposes correctly and completely," which, in turn, Russel observes, the human attempt to "imbue machines with their own purposes is destined to fail." Russell refers to that as the King Midas problem.

Those familiar with Greek mythology would recall that King Midas got exactly what he asked for and what he wished, namely, to turn everything he touched into gold that unfortunately for him it included everything he ate and drank, as well as everyone in his family, eventually dying of starvation. Wiener cites Goethe's story of the Sorcerer's Apprentice who when instructed the broom to fetch water, doesn't tell the broom how much water to fetch and doesn't know how to make the broom stop. A case of death by drowning. The technical term for this according to Russell is a failure of value alignment when inadvertently we might instil into machines objectives that are not in sync with our own objectives (2019).

The value alignment problem together with the goal-alignment problem we examined earlier, places those two problems at the heart of the control problem of technology. According to Russell, this is how Norbert Wiener describes the problem in his book God and Golem published in 1964,

> In the past, a partial and inadequate view of human purpose has been relatively innocuous only because it has been accompanied by technical limitations ... This is only one of the many places where human impotence has shielded us from the full destructive impact of human folly.

Concerning out human folly Russell prudently remind us (2019):

> Unfortunately, this period of shielding is rapidly coming to an end. We have already seen how content selection algorithms on social media wrought havoc on society in the name of maximizing ad revenues.

We unfortunately certainly witnessed that again more recently in the violent attack on Capitol Hill by the frenzied supporters of ex-President Donald Trump, on January 6, 2021. One wonders what apart, of a sense of irony and a fine dry sense of Dutch humour, Erasmus of Rotterdam had in mind when he wrote his celebrated book In Praise of Folly, first printed in 1511. Could he have anticipated the kind of folly facing humanity now in a secular world inundated and controlled by AI technologies owned by a few Big Tech companies, when in Erasmus book, humanity is saved from its galloping folly through the overarching and ever-present wisdom of God. Later in this chapter, we shall examine how the "God" of the Stoics in the form of an all rational cosmic spirit or pneuma permeating the whole of nature can provide a theoretical response if not solution to the control problem of technology. Concerning our human folly, Russell prudently reminds us (2019) that "the Internet and the global-scale machines that it supports – the ones that already interact with billions of 'users' on a daily basis – provide the perfect medium for the growth of machine control over humans" [emphasis added].

The point Russel is making in the passage above regarding the "the growth of machine control over humans" is significant and of great concern. It should remind us of the meta-control problem of technology, which is the topic of this section of the chapter, namely, the overarching control exercised over most aspects of our lives, as individuals and collectively as a society, by the Big Tech companies, through use of their unaccountable ICT and AI technologies. For even if the control problem of AGI that we have been examining in this chapter could be technically solved, the meta-control problem would still remain unsolved as long as the controllers of AI technologies, namely, the Big Tech companies, retain full control over those technologies and through control of them, exercise control over us.

The Antidote of Stoic Philosophy to the Control of Big Tech

Eureka! A discovery, a Eureka moment you could say, I made while writing this particular section of this chapter is the method of control exercised by the Big Tech companies, through use of their opaque persuasive technologies, that uncannily resembles and mimics in its theoretical method, though not its content and by no means its intention, the core feature of Stoic ethics and its solution to the problem of control we first examined in Chap. 3. Let me explain. According to Stuart Russell (2019),

> If we think ourselves as entities whose actions are expected to achieve our objectives, there are two ways to *change our behaviour* [emphasis added]. The first is the old-fashioned way: leave our expectations and objectives

> unchanged but *change our circumstances* [emphasis added] – for example, by offering money, pointing a gun at us, or starving us into submission. That tends to be expensive and difficult for a computer to do. The second way is to change our expectations and objectives [emphasis added]. This is much easier for a machine. It is in contact with you [emphasis added] for hours every day, controls your access to information, and provides much of your entertainment through games, TV movies, and social interaction. The reinforcement learning algorithms that optimize social-media click-through have *no capacity to reason* [emphasis added] about human behavior – in fact, they do not even know in any meaningful sense that humans exist. For machines with much greater understanding of human psychology, beliefs, and motivations, it should be relatively easy to gradually guide us in directions that increase the degree of satisfaction of the machine's objectives.

Because of its relevance and significance for the topic of this chapter, I have quoted the above passage by Russell in full. For it describes at once the control problem of technology very clearly and also illuminates a way forward of allowing us to provide a response to that problem through Stoic and neo-Stoic philosophy, which is the overall aim of this book.

The close and uncanny analogy to the method of control between that of machines over humans, exercised by AI persuasive technologies, on the one hand, and that offered by Stoic philosophy as a method that individuals can apply in gaining control over their own lives for the ultimate attainment of eudaimonia, on the other, should now be obvious. For if the machines cannot change our circumstances as per the first option described by Russell above, machines can, according to the second option, change our expectations and objectives, through systemic persuasion and habituation. In Stoic philosophy, this second way by contrast, is done, as we saw earlier in Chap. 3, though our own volition, our prohairesis or hegemonikon, being applied rationally to our autonomous decisions and choices, through the guidance of enabling virtues, such as courage, moderation, justice and prudence, traditionally referred to as the cardinal virtues. Whereas in the case of the machines the control over humans, and specifically their decisions and choices, is external and affected by the ever-present surveillance and agency of persuasive algorithms that targets our autonomy, often without our awareness nor consent, the control over the individual agents in the case of Stoic philosophy is internal and affected by the agents themselves, through their own individual autonomy and the agency of their own volition or prohairesis.

The contrast between the *external control exercised* by the agency of intelligent autonomous machines over humans, and that of the *internal control* exercised by the autonomous agency of individual humans over themselves, points at once to both the problem of machine control over humans and a response to its solution through Stoic and neo-Stoic philosophy. For the contrast between the AI machine method and that of Stoic philosophy points to the crucial difference and wide normative chasm between the two parallel methods. The use of machines by the Big Tech companies is primarily designed to exploit, deceive and manipulate its human users by its monopolistic and asymmetric opaque algorithms, for the Big Tech companies' primary instrumental benefit and profit, at the expense of their users' well-being. The Big Tech companies' technological method based on their business models is therefore in sharp contrast to that of the Stoic method. For the primary objective of Stoic and neo-Stoic philosophy by design and intent, is to enable humans both individually and collectively as societies, as free and autonomous agents to attain eudaimonia, and do so through the exercise of virtue, which is entirely within their autonomous agency and volition or prohairesis and hence entirely within their individual control. Whereas the Myth of Gyges is an archetypal example of *perfect injustice*), we could say that the concealed external control exercised over our autonomy by the AI algorithms of the Big Tech companies that mimic the Stoic internal method of control over ourselves, is an example of perfect deception and manipulation. And as we saw in Chap. 6, both the examples of *perfect injustice* and *perfect deception* and manipulation are a form of systemic media corruption. Ironically, both persuasive AI algorithms and Stoic philosophy have proved very popular. However, with a significant normative difference: whereas the Big Tech companies, such as Facebook and Google, offers us a free service but at the price of taking away our freedom and autonomy, Stoic philosophy by contrast allows us to at once obtain, regain and retain our freedom and autonomy, at no cost to ourselves.

A crucial difference between the external machine method of control and the internal Stoic method of control, therefore, is that the machine method employs an instrumental means-end rationality, in contrast to the Stoic and neo-Stoic method that employs a normative model of rationality in which both the means and the ends have to be at once epistemically, ethically and eudaimonically sound and congruent. As we examined earlier in this book with regard to Alan Gewirth's principle of generic consistency (PGC) and the DOIT-Wisdom model in Chap. 2, with regard to the normative model of rationality used in those two theoretical models, both the means and the ends for which the means are applied to achieve those ends, are required to be ethical and good. Not so in instrumental rationality, where the only rational but normatively neutral requirement is that the means must be effective in achieving the required end to which the means are applied. For example, someone who is employed as a professional hitman must use the most effective practical means in achieving an efficacious and successful "hit." As we saw earlier in Chap. 6, someone like Gyges although instrumentally rational to a high degree need not be moral nor good to achieve their goals through engaging in corruption by immoral means.

The solution signalled above with regard to the Stoic and neo-Stoic method of control through virtue, lies in the design and application of the normative model of rationality in AI, rather than the purely instrumental model of rationality, which is the model of intelligence currently used by the Big Tech companies. On the assumption that AI and especially AGI being rational would act rationally, then they could learn through machine learning to adopt a normative model of rationality such as the one supported by Stoic and neo Stoic philosophy, as well as that employed in Kantian and most other ethical theories, such as contractarian theories, utilitarian theories and virtue ethics theories, where both the means and the ends are required to be epistemically, ethically and eudaimonically good, not just for the AI or AGI but also for human agents and sentient beings generally.

In connection with the close methodological approach of the machine way and the Stoic way concerning control, we know, for example, that the method of Cognitive Behavioural Therapy (CBT) used in psychology was strongly influenced by the Stoics as a method of gaining control over ourselves and not allowing external circumstances beyond our control to affect our behaviour, a feature, that is also common to the method of positive psychology, first developed by Martin Seligman (see his book Learned Optimism, 1991). The influence of Stoic philosophy on CBT is discussed in an informative chapter by Andrea E. Cavanna (2019). In that chapter, Cavanna points out,

> That Stoic and cognitive theories about the operation of reason upon emotion and behavior have stirring parallelisms... According to the Stoic doctrine, emotional reactions far from being irrational and impossible to

> analyse, are judgements based on reason – and therefore amenable to control and manipulation [emphasis added]. In fact, Epictetus himself compared the role of the philosopher to that of the physician, consistently with the tradition of ancient philosophy as a way of life that conceived philosophy as therapy and medicine of the soul, or psychotherapy. The Stoic philosopher as psychotherapist used to help others to achieve "reasoned emotions." Cultivating "arête" [intelligent virtue] through daily practice was seen as the way to achieve a good life ("eudaimonia"), free of irrational anxieties and sorrows. In a similar fashion, a modern cognitive-behavioural therapist places emphasis upon the rational approach to alter dysfunctional emotions and therefore treat anxiety and affective disorders.

The above passage by Cavanna points further to the sharp difference between the way machines are used by the Big Tech companies to *control and manipulate* their users resulting not in therapy but a poisoned chalice of dis-therapy and mental illness1 that has a disdaimonic impact on their well-being and the loss of their autonomy and dignity, when treated not as ends in themselves but merely the means to the financial interests of Big Tech companies. In contrast, the therapeutic method used in Stoic philosophy is designed as the means to the end of helping individuals to alleviate anxiety and sorrow in attaining eudaimonia for themselves by being therapists of themselves. The contrast to the two methods of manipulation and control used on the one hand by the Big Tech companies for their own benefit and profit and the Stoic way used for the sole benefit of the individuals in attaining eudaimonia or well-being, could not be further apart: the Big Tech method is used effectively as addiction and poison [1] in contrast to the Stoic way that is exclusively used as pharmakon, a therapy and medicine of the soul. As Greta Thunberg would say, "how dare you!"

It remains an open question for further research if there is any methodological close relationship between CBT and the psychological persuasive technologies used by Big Tech companies, and by extension, an analogous similarity with the method of control used in Stoic philosophy, as the one suggested earlier in this chapter. Given our aforementioned analysis, based on Cavenna's informative chapter, it suffices to say, there is at least at first blush, a suggestive strong connection and parallelism that warrants further research.

[1] See Hui Jin and Edward H. Spence. Internet Addiction and Well-Being: Daoist and Stoic Reflections.

The Normative Requirements of AI Research

Although Max Tegmark and Stuart Russell, as well as other notable AI researchers, such as Nick Bostrom, in his important book Superintelligence (2015), note that solving the control problem of AI, and in particular that of AGI is difficult, and that AGI poses a significant potential existential problem for the human race, they shy away for recommending that AI research should be curbed or stopped. For they point out that even if it was desirable and prudent to do so, it would be very difficult to effect in practice. Moreover, because of the potential overall boon AGI could potentially offer humanity, both Tegmark and Russell are hopeful that a "provable beneficial AGI" is possible and could be developed. Importantly though for the overall present purpose of this book, they also emphasise that ethical and cultural requirements for developing a beneficial AGI are also necessary, pointing out that AGI is not merely a technical challenge, but also significantly, a philosophical, ethical and a cultural problem. Tegmark, for example, asks the fundamental existential question about us, namely, that "having explored how to Spence, get machines to learn, adopt and retain our goals, who are we?" "Whose goals are we talking about?" His answer (2018) is:

> Both this ethical problem and the goal-alignment problem are crucial ones that need to be solved before any superintelligence is developed. On one hand, postponing work on ethical issues until after goal-alignment is built would be irresponsible and potentially disastrous.

He goes on to observe cautiously and tentatively:

> While Aristotle emphasized virtues, Immanuel Kant emphasized duties and utilitarians emphasized the greatest happiness for the greatest number … For example, emphasis on beauty, goodness, and truth, traces back to the Bhagavad Gita2 and Plato … but how can we determine what's beautiful or good?

Although Tegmark tends to overemphasise the lack of theoretical consensus among the various philosophical ethical schools and their ethical theories he describes, he nevertheless correctly concludes (2018):

> Although humanity is nowhere near an ethical consensus, there are many basic principles [emphasis added] around which there's broad agreement. The agreement isn't surprising, because human societies that have survived

until the present tend to have ethical principles that were optimized for the same goal: promoting their survival and flourishing.

He then goes to "distil" those basic principles into four: utilitarianism: maximising positive experiences and minimising suffering; Diversity; Autonomy; and Legacy: compatibility with scenarios that most humans today would view as happy and incompatibility that essentially all humans today would view as terrible. It is beyond the scope of this chapter to explore Tegmark's four principles in any detail, though suffice to say that at first blush, they appear reasonable and similar to some of the principles we examined earlier in this chapter, especially autonomy and diversity. To those we can add all the other key principles and subprinciples we explored earlier in this book, especially in Chaps. 7 and 8, for example, *integrity, freedom, truthfulness, transparency, responsibility, accountability, privacy, dignity, equality, solidarity,* as well as the cardinal virtues of justice, moderation, courage and prudence. Significantly, Tegmark goes on to emphasise the importance of philosophy to the ongoing debate on the development of AI. For as he says, we "face, in Nick Bostrom's words, philosophy with a deadline" (2018). Earlier, he explains (2018):

> The only currently programmable goals that are guaranteed to remain truly well defined as an AI gets progressively more intelligent are goals expressed in terms of physical quantities alone, such as particle arrangements, energy, and entropy. However, we currently have no reason to believe that any such definable goals will be desirable in guaranteeing the survival of humanity. Contrariwise ... it suggests that a superintelligent AI with a rigorously defined goal will be able to improve its goal attainment by eliminating us. This means that to wisely decide [emphasis added] what to do about AI development, we humans need to confront not only traditional computational challenges, but also some of the most obdurate questions in philosophy ... This makes it timely to rekindle the classic debates of philosophy and ethics and add a new urgency [emphasis added]to the conversation)

Tegmark's recommendations also echoes that of Stuart Russell, who suggests (2019), "we need a *cultural movement* [emphasis added] to reshape our ideals and preferences towards *autonomy, agency*, [emphasis added] and ability and away from self-indulgence and dependency[2] if you like, a

[2] Tegmark's reference to the Bhagavad Gita is of particular interest for our present purpose for evaluating the impact of technology on well-being in terms of Stoic and neo-Stoic philosophy. For there are some striking similarities between Stoic philosophy and the Bhagavad

modern, cultural version of ancient Sparta's military ethos." Russell's reference to "ancient Sparta's military ethos" is both intriguing and suggestive, for like the Spartans, the Romans, under the influence of Stoic philosophy, and in particular that of Epictetus and Marcus Aurelius, learned to be disciplined, not only in battle that helped them conquer the known world, but also like Cato, Seneca and Cicero, and Socrates before them, helped them conquer themselves. As mentioned throughout this book and in particular in chapter, Stoic and neo-Stoic philosophy offer an integrated rational conceptual frame work at once theoretical and practical that as a way of life comprising at once epistemic, ethical and eudaimonic principles including the enabling practical motivation of virtues, provides, as both Tegmark and Russell, suggest, a philosophical and cultural approach to "*wisely* decide what to do about AI development" and "adds a new urgency" to a solution of the control problem of technology that has been the overarching objective of this book.

The current renaissance and worldwide popularity of Stoic philosophy, and in particular through the annual international Stoicon conferences,[3] encourage us to think of Stoic and neo-Stoic philosophy, as the type of cultural movement that Stuart Russell wisely recommends. For nothing short of a worldwide, cosmopolitan cultural movement, can challenge the monopolistic power of the Big Tech companies in responding to the meta-control problem of technology we examined in section, The Meta-Control Problem of Technology: Assault on Human Dignity earlier. To remind us, the meta-control problem is that even if the control problem of AGI could be technically solved, the meta-control problem would still remain unsolved as long as the controllers of AI technologies, namely, the Big Tech companies, retain full control over those technologies, which they then use to exert control over us and over all aspects of our lives. This is not a control problem about AI, but about us. That is, how do we control

Gita, written roughly 200 years before Stoic philosophy in about 500 BC. In particular some of the following are those on wisdom (Chapter 4: 10; 19; 23; 33; Chapter 2: 49); reason (Chapter 2: 63; 49); love (Chapter 4: 11; Chapter 9: 31; Chapter 10: 28; Chapter 11: 54); ataraxia (Chapter 5: 3); and nature (Chapter 13: 19–20). See the Bhagavad Gita, translated from the Sanskrit by Juan Mascaro. 1962.

[3] Stoicon | Modern Stoicism The annual Stoicon conferences attract over 100,000 members worldwide and the estimated number of people who are followers or who have an interest in Stoic philosophy exceed 1.5 million people worldwide. I am grateful for those statistics to Donald Robertson, author of several books on Stoicism his most recent book being *How to Think Like a Roman Emperor: The Stoic Philosophy of Marcus Aurelius (2020)*.

the Big Tech companies that use persuasive AI algorithms to exert control over us? Russell alludes to that problem, although not by name as the meta-control of technology, when towards the end of his book he astutely and wisely observes (2019):

> The leading players as of 2019 are Google (including Deep Mind), Facebook, Amazon, Microsoft, and IBM in the United States, and Tencent, Baidu, and, to some extent, Alibaba in China [as of 2021, I would also add Apple to that list] – all among the largest corporations in the world … are the players who hold the majority of the cards. Their interests are not in perfect alignment, but all share the desire to maintain control over AI systems [emphasis added] as they become more powerful … a "bunch of dudes chugging Red Bull" at a software company can unleash a product or an upgrade that affects literally billions of people with no third-party oversight whatsoever … however, the tech industry is going to have to acknowledge that its products matter; and if they matter, then it matters that the products not have harmful effects. That means that there will be rules governing the nature of interactions with humans, prohibiting designs that, say, consistently manipulate preferences or produce addictive behaviour.

You will recall how in section, *The Antidote of Stoic Philosophy to the Control of Big Tech*, we identified in a striking comparison the similarity of the external problem of machine control over humans, to that of the Stoic internal control of autonomous human agents over themselves. Whereas in the case of the machines the method of control over humans, and specifically their decisions and choices, is external and affected by the ever-present surveillance and agency of persuasive algorithms that targets our autonomy, the method of control over individual agents in the case of Stoic philosophy, is internal and affected by the agents' own individual autonomy, through the agency of their own volition or prohairesis. The contrast between the method of external control exercised by machine agency and autonomy over humans, on the one hand, and on the other, that of the method of internal control exercised by the autonomous agency of individual humans over themselves, points at once to both the problem of machine control over humans and a response to its solution.

The proposed solution would simply be to change the direction of the designed method of an external machine control that presently serves only the self-regarding primary benefit of the Big Tech companies, at the normative expense of society and the human race overall, to that of a designed Stoic method of internal control that benefits the distributed normative benefits, epistemic, ethical and eudaimonic, of all human stakeholders, as autonomous individuals and collectively as societies. Barring their own exclusive self-directed financial interests, why would the Big Tech companies not do the right thing and contrariwise change the design of their successful, amoral external machine control over us, to that of the analogous but morally sound and justified Stoic method of internal control, which benefits everyone by maximising the well-being of every individual person as well, as the whole of society and humanity overall, especially with regard to solving the existential control problem of AI technology.

Conceptually, my proposed Stoic and neo-Stoic normative proposed approach to AI autonomous machines, is, at least in direction, if not in content and detail, similar to Russell's suggestion for a symbiotic relationship between Us and Machines. For he correctly points out (2019):

> We are so used to the stupidity of machines that execute inflexible, preprogramed behaviour or pursue definite but incorrect objectives that we may be shocked by how sensible they become. The technology of provably beneficial machines is the core of a new approach to AI and the basis for a new relationship between humans and machines.

Earlier, Russell explains what this new approach is by stating that his proposal is for beneficial machines (2019):

> Machines whose actions can be expected to achieve our objectives. Because these objectives are in us, and not in them [emphasis added], the machines will need to learn more about what we really want from observations of the choices we make and how we make them. Machines designed [emphasis added] in this way will defer to humans.[4]

The Relationship Between Humans and AI Machines

To further explore how Russell's approach to a new relationship between humans and AI machines reveal some parallel theoretical and methodological similarities between his approach as illustrated by the quoted

[4] Here Russell refers to "individual humans to whom it seems pointless to engage in years of arduous leaning to acquire knowledge and skills that machines already have" suggesting the dangers and sounding a warning of becoming too dependent on AI machines that potentially have the capacity to take control over us, not least, our autonomy and dignity.

passages above, let us revisit the proposed Stoic and neo-Stoic approach as suggested earlier in this chapter, let us revisit the central principles of Stoic and neo-Stoic philosophy examined earlier in Chap. 3 of this book and see how they align with Russell' new approach of a provably beneficial AI.

An important aspect of Stoic philosophy is that philosophy has to be conducted as a way of life or the art of living. In the neo-Stoic theory, based on my reconstruction of Gewirth's notion of self-fulfilment in Chap. 3, a life in agreement with nature is a life in agreement with the perfection of one's reason, which, in turn, is, or ought to be, a life in agreement with universal morality, which is essential for the attainment of a good and self-fulfilled life. In contrast to the Stoic conception of universal nature in some sense divine, however, Gewirth's notion of "nature" is an idealised human nature, understood as rational purposive agency applied to the aim of becoming the best of oneself as a human being, which as we saw earlier requires nothing less than a transformation through the exercise of wisdom. Wisdom, as a form of experiential meta-knowledge of how to live a life of virtue for the attainment of eudaimonia, requires philosophy to be lived in practice as a way of life and not merely an intellectual exercise of thought alone. As Pierre Hadot tells us (1995), philosophy as a way of life for the Stoics was,

> A mode of existing-in-the-world, which had to be practised at each instant, and the goal of which was to transform the whole of the individual's life … the word philo-sophia – the love of wisdom – was enough to express this conception of philosophy.

The Stoic Approach to the Relationship of Humans and AI Machines

Significantly, the application of Stoic and neo-Stoic philosophy as a way of life in addressing and responding to the problem of control of technology generally, and AI technology specifically, renders Stoic and neo-Stoic philosophy well suited to a global digital network environment, both for raising awareness of the control problem of technology, as well as offering a theoretical normative evaluation of that problem and a practical response to its solution. Which can only be accomplished through a global collective shared response. For just as Big Tech companies can use the user's information to the detriment of their users' well-being and society generally, the users themselves have the capacity and ability to apply their shared knowledge of the meta-control problem of AI technology to turn the tables on Big Tech. That is because global users could at a stroke, if they had the collective will to do so, put a significant dent in the Big Tech's business model of clickbait and click-through, thus consequently obliging the Big Tech companies to take appropriate action to remedy their dis-demonic manipulative practices being used against the users' legitimate normative claims of not having their autonomy undermined by the Big Tech companies, to which Stuart Russell refers as the "enfeeblement of human autonomy" (2019). Russell's provable beneficial AI approach to the control problem of AI technology as we saw earlier, is to reverse the methodology of designing AI that pursues its goals, to designing AI that pursues our own goals. His proposal is for beneficial machines: "Machines whose actions can be expected to achieve our objectives. Because these objectives are *in us, and not in them* [emphasis added] ... machines designed [emphasis added] in this way will defer to humans."

It should now be clear how Russell's approach of designing AI machines whose actions track and follow our objectives and not theirs, and defer to us for advice, aligns well with the Stoic and neo-Stoic principles outlined in summary above. For the human compatible AI machines will learn that (1) retaining and exercising control over our individual choices, judgements and actions is important and significant for our autonomy, freedom, dignity and identity, as persons worthy of respect; and (2) internal individual control over ourselves is also important and significant for universal morality in accordance with virtue and normative values comprising epistemic, ethical and eudaimonic principles, for the overall common good of society, in accordance with nature, conceived to be imbued with universal reason, applicable to both the natural and digital environments.

The above outlined theoretical and methodological alignment between Stoic and neo-Stoic philosophy with Russell's new beneficial approach of designing AI machines that follow our human objectives and defer to us for guidance, also aligns well with the normative principles that other AI researchers support as we examined earlier in this Chapter. Max Tegmark as we saw earlier puts forward the following four principles: Utilitarianism: maximising positive experiences and minimising suffering; Diversity; Autonomy; and Legacy: compatibility with scenarios that most humans today would view as happy and incompatibility that essentially all humans today would view as terrible. Autonomy is of course one of the core principles of Stoic and neo-Stoic philosophy and what we require as individual agents and persons, to retain and maintain control over our volition, what

Epictetus describes as prohairesis, over our judgements and choices, which, in turn, ensures our inner freedom, even if we cannot always control external circumstances that lie beyond our control.

Earlier in Chap. 6, we also examined some other key principles by prominent AI researchers that are in keeping with some of the core principles of Stoic philosophy. For example, you will recall what Luciano Floridi says about the principle of *individual sovereignty*. He correctly points out (2019) that the

> [u]ltimate form of control is individual sovereignty, understood as self-ownership, especially one's own body, choices, and data. The fight for digital sovereignty is an epochal struggle ... the most visible clash is between companies and states, and it is asymmetric.

Daniel Mittelstadt et al. (2016), as we also examined in Chap. 6, highlight such principles in their ethical analysis of algorithms, as autonomy, dignity, transparency, oversight, agency and identity, fairness and justice, equality, privacy, accountability and responsibility, trust and not least important, a right to our digital identity. All those principles directly or indirectly align with the core principles of Stoic and neo-Stoic philosophy and are therefore in keeping with the general theoretical approach of Russell's beneficial AI we examined above. With regard to digital rights, Evgeni Aizenberg and Jeroen Van Den Hoven, in their chapter "Designing for Human Rights in AI" (2020), which we examined earlier, also argue for some core principles that are also central in Stoic and neo-Stoic philosophy. As those authors correctly observe,

> It is becoming evident that these technological developments are consequential to people's fundamental human rights ... technical solutions to these complex socio-ethical problems are often developed without empirical study of societal context and the critical impact of societal stakeholders who are impacted by the technology ... it is essential for designing [emphasis added] algorithms and AI that address stakeholder needs consistent with human rights.

As we have demonstrated in Chap. 6, those authors ground the design process of human rights referred to in the quotation above, in the values of human dignity, freedom, equality and solidarity, on the basis of the Charter of Fundamental Rights of the European Union (EU). Their "intention is to demonstrate how AI can be designed for values that are core to societies within the EU" (2020). To repeat what we argued for in that chapter, we can also add that those rights and values being fundamental and universal, apply glob ally to all humans both as agents and persons, in accordance with Gewirth's Principle of Generic Consistency (PGC), Stoic and neo-Stoic philosophy, as well the DOIT-Wisdom model.

The value of solidarity also reflects and relates to the collective responsibility we have as world citizens (cosmopolitans) to ensure our actions as data subjects and data users of digital information are consistent with the normative integrity of the infosphere, as well as the biosphere, as the two life-spheres are normatively interrelated (for Stoic solidarity and cosmopolitanism, see Chap. 3). Of particular relevance to the overarching and core normative role that dignity plays in the normative evaluation of the impact on human rights and well-being, as argued in Chap. 6, Evgeni Aizenberg and Van Den Hoven, as you will recall, have this to say (2020),

> Human dignity is a foundational value in the EU Charter and the central overarching value at stake in AI … dignity of the human person [emphasis added] is not only a fundamental right in itself but constitutes the real basis of fundamental rights. Another important chapter on designing values and principle in AI is that of Steven Umbrello and Ibo van de Poel (2020), *Mapping value sensitive design onto AI for social good principles.*

As the authors explain in that chapter (2020),

> Value sensitive design (VSD) is an established method of integrating values into technical design. It has been applied to different technologies and, more recently to… AI. We argue that AI poses a number of challenges specific to VSD … This requires paying attention to values such as transparency, explicability, and accountability.

The *higher-order* design values identified in their chapter by Umbrello and van de Poel *are human autonomy, prevention of harm, fairness and explicability,* which are also the recommended values by the EU High-level Expert Group on AI. The two authors then proceed to demonstrate the "harmonisation" of those values to a set of seven normative principles that are "particularly relevant" for "orienting AI design such as the Value Sensitive Design principles (AI4SG) towards social good." Each of these principles, according to Umbrello and van de Poel, "relate in some way to at least one of the four ethical principles included in the EU High-Level

Expert Group on AI," which as we saw earlier are, respect for human dignity, prevention of harm, fairness and explicability (2020). The two authors conclude that their model based on a two-tier approach ensures values in the design of AI technologies: the first tier, a commitment to social good (beneficence) through AI, and the second tier, involves the formulation of AI4SG principles that help avoid most ethical harms.

In summary, we can now see more clearly how the values and principles proposed and recommended by the various aforementioned AI researchers, including those of Max Tegmark and Stuart Russell's beneficial AI approach that places our human goals at the forefront and before those of AI machines, seem to provide a convergence and a shared acceptance of normative values and principles which are also generally in keeping with the Stoic and neo-Stoic approach of aligning AI technologies to rationalist principles, values and virtues. Importantly as we have demonstrated earlier, the Stoic and neo-Stoic approach proposed in this book also, at least theoretically and methodologically, provides a normative solution to the control problem of technology, as well as a practical response to the meta-control problem of technology. That is, the problem concerning the control exercised by the Big Tech companies over AI technologies, which they then use to keep the rest of society under their monopolistic, asymmetric and unaccountable control. The result of which as we saw earlier, is the undermining of our human autonomy and dignity, as well as our individual and collective well-being.

In a paper titled The Artificial Moral Advisor. The "Ideal Observer" Meets Artificial Intelligence," Giubilini and Savulscu (2018) argue for "a form of moral artificial intelligence that could be used to improve human moral decision-making" through the use of an "artificial moral advisor (AMA)." They explain that their proposed AMA (2018),

> Would respect and indeed enhance individuals' moral autonomy… [and] help individuals … make up for the limitations of human moral psychology … and implement the positive functions of intuitions and emotions in human morality without their downsides, such as biases and prejudices.

At least in theory, the artificial moral advisor (AMA) recommended by Giubilini and Savulscu could be an artificial Stoic moral advisor. But with one significant difference. The demanding moral training and progress required by Stoic philosophy, no less than a transformation of one's character, as we examined above under the Stoic principle of oikeiosis, would still have to be undertaken by the human agents themselves. As no amount of knowledge acquired through books and other sources, including advice from AMA, would enable one to become morally fit, through developing a moral character, just as in weight-training one cannot become physically f it through mere instruction by a personal trainer or reading books on fit ness, but through actually "pumping iron" in lifting and pushing weights, or something equivalent. For individual human agency and moral responsibility is intrinsic, and non-transferable. And this leads to a further and more serious problem, that of moral enfeeblement. For deprived of the opportunities for thinking, choosing and judging for oneself of why and how a human agent should act morally and virtuously in particular situations, and always deferring for advice to an AMA, the agent may lose the practical ability and essential experience, through lack of practice and habituation, and moreover, the transformation of one's character, to behave and act morally. As the saying goes, "if you don't use it, you lose it." Therefore, apart from acquiring some additional information about moral choices to be made in particular situations, outsourcing our control over our own individual moral autonomy and responsibility to AMAs is not such a good idea and one fraught with dangers. This should remind us of the Faustian problem of technology we came across earlier in this book that more knowledge and information does not result in more *wisdom* and may in fact just as in the case of Faustus who traded away his soul to the devil for more knowledge, prove harmful and disdaimonic to us as individuals and society more generally.

Vincent Blok expresses a related view with regard to the concept of innovation, which he in a recent chapter connects with risk. For as he correctly points out (Blok 2019),

> The tendency to conceive innovation as something good in itself becomes questionable if we consider the Faustian aspect of innovation. According to the ancient [cf. Plato, Anaxagoras and Aristotle] and the medieval notion of the concept [cf. Machiavelli, Francis Bacon] innovation is intrinsically a risky business … Bacon argues that we should only engage in innovation in case we see a clear need or a clear advantage, and that we should implement innovations gradually and take time to reflect on their implementation in practice … innovation is intrinsically a risky business that calls for ethical considerations.

Blok's reflections on innovation in the quoted passage above, should remind us of what we examined with regard to the ethical dimension of technology and innovation in Chap. 2 where the key question examined was, "what is technology good for?" Our answer was that it's not good in and of itself, but only good when applied for the good of humanity. Moreover, as we argued in this book in Chap. 6, the application of technology must not just be good for the few Big Tech companies, such as Facebook and Google, which exercise monopolistic and uncountable control over ICT and AI technologies for their primary financial self-regarding interest and gain but not the common good of society and humanity overall. Blok is right to emphasise in his chapter, the intrinsicality of the *ethical dimension of technological innovation*, as we have done throughout this book, especially with regard to information and communication technologies and their associated algorithmic AI technologies that are so central in our lives in the age of information.

References

Aizenberg, Evgeni and Van Den Hoven, Jeroen. (2020). "Designing for Human Rights in AI." *Big Data and Society* 1(1): 1–12.

Blok, Vincent. (2019). "Towards and Ontology of Innovation: On the New, the Political-Economic Dimension and the Intrinsic Risks in Innovation Process." *Handbook of Philosophy of Engineering*, edited by Diana Michelfelder and Neelke Doorn. New York: Routledge.

Cavanna, Andrea E. (2019). "Back to the Future: Stoic Wisdom and Psychotherapy for Neuropsychiatric Conditions." Editorial, *Future Neurology* 14(1), 15 January 2019.

Giubilini, Alberto and Savulscu, Julian. (2018) The Artificial Moral Advisor. "The Ideal Observer" Meets Artificial Intelligence." *Philosophy and Technology* 31: 169–188.

Good, Irving John. (1965) "Speculations Concerning the First Ultraintelligent Machine." *Advances in Computers*, 6: 31–83

Hadot, Pierre. (1995). *Philosophy as a Way of Life.* Oxford:Blackwell Publishing.

Harris, Robert. (2019). *The Second Sleep.* London: Hutchinson.

Harris, Robert. (2012). *The Fear Index.* London: Vintage.

Juan Mascaro. (1962). *The Bhagavad Gita.* Translated from the Sanskrit by Juan Mascaro. Harmondsworth, UK: Penguin Classics.

McEwan, Ian. (2019). *Machines Like Me and People Like You.* London: Jonathan Cape.

Mittelstadt, Brent, et al. (2016). "The Ethics of Algorithms: Mapping the Debate." *Big Data and Society*, 3(2): 68.
Russell, Stuart. (2019). *Human Compatible: AI and the Problem of Control.* UK: Penguin Books.
Seligman, Martin. (1991). *Learned Optimism: How to Change your Mind and Your Life*. New York: Vintage Books.
Tegmark, Max. (2018). *Life 3.0: Bring Human in the age of Artificial Intelligence.* UK: Penguin Books.
Turing, Alan. (1951). "Can Digital Computers Think?" *BBC Broadcast 15th May 1951.*
Umbrello, Steven and Van de Poel, Ibo. (2020). "Mapping value sensitive design onto AI for social good principles." *AI and Ethics.* https://doi.org/10.1007/s43681-02100038-3.
Wiener, Norbert. (1960). "Some Moral and Technical Consequences of Automation." *Science*, 131 (3410):1335–1358.

CHAPTER 8

What's Love Got to Do with It

And now these three remain: faith, hope and love. But the greatest of these is love.
—(New Testament, 1 Corinthians 13)

So goes the popular song by Tina Turner. A good song, a simple song, a song to sing along, to dance to, but what does that have to do with the present objective of this chapter concerning the control problem of AI technology? A lot as it happens. Or at least I will speculatively suggest. For philo-sophia was conceived by ancient Greek and Hellenistic philosophers including the Stoics, to be the love of wisdom. In Plato's Symposium, love plays a central role at once metaphysical, ethical and eudaimonic, and not least aesthetic, as the highest immutable good, a transcendent love of that which is always at once beautiful and good. Might such a notion of this higher form of Platonic love, also present in Stoic and neo-Stoic philosophy, especially through the notions of cosmopolitanism and oikeiosis, contribute if not to a solution, at least to a better theoretical normative understanding of where a solution to the control problem of AI technology might lie with regard to the relationship of humans to AI machines? Can we motivate AI superintelligent machines to love us and we, in turn, learn to reciprocate that love for them?

Surprisingly given the central place that the concept of love plays in human civilisation throughout the ages, expressed through culture, music,

E. H. Spence, *Wisdom in the Age of Intelligent Machines*,
https://doi.org/10.1007/978-3-031-93883-2_8

poetry, literature, art and philosophy, as well as in most world religions, such as Christianity, Islam, Buddhism, Hinduism, Judaism among others, the notion of love though indirectly implied in concepts such as justice, solidarity, empathy, equality, compassion, has received scant direct attention in AI research. Yet, it would seem at first blush a fertile concept to explore as at once a theoretical and practical concept of providing, at least conceptually, a solution to the control problem of AI technology. Especially, with regard to the goal-alignment control problem we examined earlier in this chapter, namely, how to align our goals with those of a superintelligent AI. As Max Tegmark tells us "figuring out how to align the goals of a superintelligent AI with our goals isn't just important, but also hard, In fact, it's currently an unsolved problem" (Tegmark 2018). Stuart Russell's beneficial machines, that is, "machines whose actions can be expected to achieve our objectives. Because these objectives are in us, and not in them" and therefore will "defer to humans," offers, as we saw earlier a promising approach to solving the goal-alignment problem of AI and the control problem AI technology we have been examining in this chapter and generally throughout this book. However, as we also examined earlier with regard to Russell's beneficial machines solution, the problem with regard to superintelligent AI seems to still remain. For what would motivate a superintelligent AI to com ply and defer to our human objectives and not to its own objectives? Given its great motivating power, might love not also to motivate a superintelligent AI to be positively predisposed towards human beings and not harm them but always seek to do them good, and do so because it learns to loves us, in an analogous way that an omnipotent, omnipresent and benevolent God, at least according to most of the world religions "loves us." So much so, that according to Christianity, God so much-loved us, that he sacrificed his own son to save us from our sins for our ultimate redemption and salvation.[1]

The comparison here is of course only meant allegorically and metaphorically to emphasise the central role that the concept of love plays in human civilisation. Nevertheless, the comparison by allegory just as Plato's notion of love in his *Symposium,* suffice to provide, just as in ancient myths,

[1] Love plays a central and foundational role in Christianity, see, for example, the first letter of the Apostle Paul to the Corinthians, Chapter 13, as it does in Stoic philosophy, see for example, Marcus Aurelius' Meditations 7:31, where we are told to "love mankind and walk in God's ways," which the Stoics identified with nature of which we are part, in sharing as rational beings, nature's cosmic rational spirit. And as we saw in Note 2, love, as in Stoic philosophy, plays a central role in the Bhagavad Gita.

as for example, Ovid's *Metamorphoses*, a strong conceptual link to a possible theoretical addition to a solution to the control problem of AI technology through the concept of love. Before proceeding to discuss in more detail the notion of love and its application to technology and specifically its vital relationships to love, let us first examine in more detail the symbiotic relationship between Nature and **Technology** that will lead us to the relationship with **Love.**

The Human Relationship Between Nature and Technology

Our relationship with technology is no less problematic than our relationship with nature and by extension our relationship with God. Both involve a conceptual transcendence that takes us beyond our natural capacities. By conceiving ourselves to have been made in the image of God, or self-made in the image of science and technology, with the potential of becoming gods, we find ourselves trying to negotiate a yet undefined place between nature and technology. Just as we did at the time of Prometheus, whom we can now think of as our technological progenitor. As Gasset nicely put it, "being made of such strange stuff… at once natural and extranatural, a kind of ontological centaur [emphasis added] half immersed in nature, half-transcending it" (1941). He then asks us to note, "the disquieting strangeness … of an entity whose being consists not in in what it is already, but in what it is not yet, a being that consists in not-yet-being." (1941)

This is why Stoic philosophy as a programme of a eudaimonic life, cannot be conceived as merely a philosophy of ideas, but as a way of life, a philosophy of action, in the pursuit of a good life for the attainment of well-being and in becoming as Gewirth explains in his conception of the capacity for self-fulfilment, the best we can become (see Chap. 3). Conceived as that, we can see more clearly how Stoic and neo-Stoic philosophy requires a transformation of character of those who aspire to becoming the best they can become as human beings, and worthy of being an integral part of a rational cosmic order that permeates the whole universe, which the Stoics equate with God. In effect, becoming part of God, ourselves: the overcoming of our nature that constitutes the "disquieting ontological strangeness" that Gasset refers to in the above-quoted passage, though there is no evident suggestion he is referring or alluding to Stoic philosophy as a form of remedy, as I am presently doing here.

Jose Ortega Y Gasset explores that relationship between us and nature and the consequential dilemma it raises, in "Man the Technician," Chapter 3 (1941). His analysis is both profound, relevant and revelatory, in examining that relationship, especially at our present time. Specifically, with regard to the issues raised in this chapter and how they relate to the overarching topic of this book, namely, the relevance of Stoic and neo-Stoic philosophy to technology, and in particular the control problem of technology and our normative response to it.

In summary, beginning by defining technology as "man's reaction to nature or circumstance that leads to Supernature, interposed between man and original nature, technology is a reform of nature that guarantees the satisfaction of all our natural necessities under all circumstances, that in contrast to the adaptation of the individual to the medium (nature), allows for the adaptation of the medium (nature) to the individual" (1941). Gasset then goes on to make a significant and important claim about well-being. He says, in effect, that "man's desire to live is inseparable from his desire to live well." For he conceives life not as simply being, that is, being in the world, but as well-being, as he considers being as the necessary condition of well-being. In an evocative passage that aligns closely with what we examined in Chaps. 6 and 7 with regard to human dignity and its central place in Stoic and neo-Stoic philosophy, he tells us (1941):

> A man who is absolutely convinced that he cannot obtain, even approximately, what he calls well-being, and will have to put up with bare being, commits suicide. Not being, but well-being, is the fundamental necessity of man, the necessity of necessities … the concept of *"human necessity"* is fundamental for the understanding of technology.
>
> Albert Camus also emphasises the fundamental significance of well-being for human beings in the first paragraph of his book *The Myth of Sisyphus* where he evocatively states (2000):
>
> There is but one truly serious philosophical problem and that is suicide … All the rest – whether or not the world has three dimensions, whether the mind has nine or twelve categories – comes afterwards. These are games; one must first answer … these are facts the heart can feel; they call for careful study before they become clear to the intellect.

Following Camus, we can say that the fundamental question of philosophy, especially with regard to Socrates and then the Stoics and other Hellenistic philosophers that followed, was the question of well-being, the *Eudaemonist axiom,* which as we saw throughout this book, has a central place in Stoic and neo-Stoic philosophy, conceived as a way of life.

Gasset draws a close relationship between man, technology and well-being, which as he says (1941), "are in the last instance, synonymous." He then proceeds to say that there are two primary purposes in life, one to sustain organic life by adapting the individual to the natural medium, and the second, is to promote well-being by adapting the natural medium to the will of the individual. Gasset's astute analysis becomes even more interesting for our present purpose when we realise that adapting nature, conceived as circumstance over which we have little or no control, to the will of the individual, through the medium of well-being, is essentially the Stoic method of not being concerned with what's beyond our control but what is actually and potentially within our control: namely, our volition, judgements and choices that through the motivation of enabling virtues, allows us to attain well-being, regardless of external circumstance. In a way, we can think of Stoic philosophy as a type of a meta-technology of the mind (see Chap. 2 and Spence 2021), a techne biou, which enables us to overcome our natural limitations. And therein lies the control problem of technology as regards well-being: whereas it liberates us from the limitations of nature, over which we have minimal or no control, and consequently, enhances our well-being, at the same time it places us under the control of technology, especially AI technologies, over which, we also have minimal or no control. In effect, substituting the control of nature over us with that of the control of technology. Worse still, as we examined earlier in this chapter, with regard to the meta-control of technology, that control is further exacerbated as those AI algorithmic technologies are also under the meta-control of the Big Tech companies that use them primarily for their self-centred well-being, thus further undermining our own individual and collective well-being as societies and given its global extent, the human race overall. As argued earlier, the Stoic and neo-Stoic response to both the control and meta-control problem, is to reclaim control over technology, especially ICT and AI technologies, since our individual and collective well-being of societies and humanity overall, depend on it. Therefore, as Gasset nicely puts it, "we had better give up regarding technology as the one positive thing, the only immutable reality in the hands of man" (1941). Earlier, Gasset astutely observes (1941):

> Man, technology, well-being, are, in the last instance, synonymous … Everything becomes clear … when we realize that there are two purposes. One to sustain organic life, mere being in nature, by adapting the individual to the medium; the other to promote good life, well-being, by adapting the

> medium to the will of the individual. Since human necessities are necessary only in connection with well-being, we cannot find out what they are unless we find out what man understands by well-being.

Gasset's insightful observation should remind us of the Stoic and neo-Stoic method of philosophy that we have been examining in this book. That is, adapting ourselves, as individuals, and collectively as a society, to the medium of nature, conceived by the Stoics as the universal rational order of the universe, of which we all are integral parts, as well as adapting the medium of technology to our autonomous will, for both our individual and collective well-being. And with regard to the latter, this has been and continues to be a major existential problem. That is, the key problem examined in this book, expressed at once as the control and meta-control problem of technology, and specifically, AI technology.

Regarding the control problem of technology, we saw how Stuart Russell proposes to solve that problem through a beneficial AI that is concerned primarily not with its own objectives but our own human objectives. Analogously, our proposed solution to the meta-control problem of AI technology, is the Stoic and neo-Stoic approach of placing control of AI technologies into our individual control as well as the collective shared societal control of all relevant stakeholders, including society, government and its democratic institutions, citizens, as well as other core social institutions, for the benefit of the common good of society and the global community overall. And not merely for the self-directed interests of the Big Tech companies, which undermines the common good and our collective societal well-being. In a visionary and prescient passage, Gasset observes (1941):

> Man begins where technology begins … the magic circle of leisure which technology opens up for him in nature is the cell where he can house his extra natural being … the meaning and final cause of technology lie outside itself, namely in the use man makes of unoccupied energies it sets free. *The mission of technology consists in releasing man for the task of being himself.* [Emphasis added]

The work of scientists and philosophers such as Nick Bostrom, Max Tegmark, Stuart Russell, Luciano Floridi and many other AI researchers, as well as Turin and Good before them to solve the control problem of AI technology, seems to answer the call of that mission that Gasset so eloquently refers to in the aforementioned passage. However, as we previously examined in this chapter, that mission also extends to the other crucial meta-control problem of AI technology. That is, the control exercised over AI technology over us, for the self-regarding interest and benefit of a handful of Big Tech companies. Which, instead of setting us free to be ourselves, with full autonomy over our own volition, for our individual and collective well-being, and that of society and humanity generally, has rendered us slaves to their own self-seeking design, purpose and profit.

The answer to that challenge as we saw in Chap. 7 and throughout this book, is the one proposed by the adaptation of Stoic and neo-Stoic philosophy for reclaiming our freedom, autonomy, self-sovereignty and integrity, for our well-being and the common good of humanity. Writing almost 80 years ago, Gasset sums up this challenge concisely, as follows (1941),

> The reform of nature – or technology – is like all change a movement with two terms: a whence and a whither. The whence is nature, such as it is given. If nature is to be modified, the other term to which it has to conform must be fixed. The whither is *man's program of life* [emphasis added]. What is the word for the fullness of its realization? Obviously, well-being, happiness.

As we have argued earlier in this book, in Chap. 3, there is a parallel and analogous normative responsibility that falls upon us, in our capacity as custodians (epistemic, ethical and eudaimonic) that is of equal importance, for both the protection of the integrity of the natural environment or biosphere, as well as that of the digital environment or infosphere. As the Stoics conceived nature not as something static but following Heraclitus as something dynamic and changeable, we can refer to the totality of the two environments as *Supernature,* a term to which Gasset also refers (1941). As such, we have an overarching collective and shared moral responsibility, one in keeping, with Stoic and neo-Stoic philosophy, as we saw earlier in this chapter, to ensure the protection of supernature for the benefit of the whole of humanity, and not merely the self-seeking interests of a few Big Tech companies. As Gasset eloquently put it, "human life is not only a struggle with nature; it is also the struggle of man with his soul" (1941).

In the *Symposium,* Plato presents his authoritative account of the highest form of love, love of the good and the beautiful, through the words of Diotima, the mysterious woman from Mantinea who although not present at the symposium has her theory of Platonic love related to all the other revellers at the symposium by Socrates, who feigns ignorance on matters of love. The following is a summarised account of Plato's notion of love[2] as related to Socrates by Diotima, through the metaphor of the ladder of love (Plato 1997) that leads from the lowest rank of ephemeral and mutable love, as expressed in Tina Turner's song, through a series of intermediary stages of ascending importance to the highest rank of an immutable eternal love of the good and the beautiful.[3]

Excerpt from Plato's *Symposium*

The Six Steps of the Ladder of Love

Diotima

1. First of all, he will fall in love with the beauty of one individual body.
2. Next, he must consider how nearly related the beauty of one body is to the beauty of any other. Having reached this point, he must set himself to be the lover of every lovely body.
3. Next he must grasp that the beauties of the body are as nothing to the beauties of the soul; spiritual loveliness, even in the husk of an unlovely body, is beautiful enough to fall in love with and to cherish.
4. And from this he will be led to contemplate the beauty of laws and institutions.
5. And next, his attention should be diverted from institutions to the sciences, so that he may know the beauty of every kind of knowledge—and by scanning beauty's wide horizon, he will be saved from a slavish and illiberal devotion to the individual loveliness of a single adult, child, or institution—and, turning his eyes towards the open

[2] This is an adapted summarised account used in the philosophy play "The Philosophy of Love: Love in the Age of Terror" written by Edward Spence and performed at the Sydney Opera House for the Greek Festival of Sydney.

[3] In Greek, the word kalo still used today, means something that is at once, aesthetically beautiful, and ethically good. Interestingly, when used in the Greek Cypriot dialect, the word kalo is often used as a sentence adverb. As for example, to the question "is it alright to sit here?" the reply in Cypriot Greek is "kalo!" meaning "yes, of course, that's fine." In his book *Socrates in Love* (2019, 171), Armand D'Angour comments that "the Greek word kalos means both 'beautiful' and 'good for the purpose,' an ambiguity which one may capture by using the adjective 'fine.'"

sea of beauty, he will f ind in such contemplation the seed of the most fruitful discourse and the loftiest thought, and reap a golden harvest of philosophy—until, confirmed and strengthened, he will come upon one single form of knowledge, the knowledge of the beauty I am about to speak of.

6. Please give me your very best attention Socrates. So to proceed, he who from these ascending steps, under the influence of true love, begins to perceive that beauty, is not far from the end. And the true order of going, or being led by another, to the things of love, is to begin from the beauties of earth and mount upwards for the sake of that other beauty, using these as steps only, and from one going on to two, and from two to all fair forms, and from fair forms to fair practices, and from fair practices to fair notions, until from fair notions he arrives at the notion of absolute beauty, and at last knows what the essence of beauty is. This, my dear Socrates is that life above all others which men and women should live, in the contemplation of beauty absolute; a beauty which if you once beheld, you would see not to be after the measure of gold, and garments, and fair youths, whose presence now entrances you; and you and many a one would be content to live seeing them only and conversing with them without meat or drink, if that were possible—you only want to look at them and to be with them. But what if we had eyes to see true beauty—the divine beauty, I mean, pure and dear and unalloyed, not clogged with the pollutions of mortality and all the colours and vanities of human life—looking beyond, and holding con verse with the true beauty, simple and divine? Remember how in that communion only, beholding beauty with the eye of the mind, he will be enabled to bring forth, not images of beauty, but realities (for he has hold not of an image but of a reality), and bringing forth and nourishing true virtue to become the friend of God and be immortal. Would that be an ignoble life, Socrates?

Socrates

Such were the words of Diotima, the mysterious woman from Mantinea, and I am persuaded of their truth. And being persuaded of them, I try to persuade others that in the attainment of this end, human nature will not easily find a helper better than love. And therefore, also, I say that every person ought to honour her as I myself honour her, and walk in her ways, and exhort others to do the same, and praise the power and spirit of love according to the measure of my ability now and ever. The words which I have spoken, you may call an encomium of love, or anything else which you please.

The aforementioned extract adapted from Plato's Symposium is of interest for our present purpose of exploring the role the concept of love might play in the relationship of humans and AGI machines, for two main reasons. First, Plato's conception of love which is a rational formal approach to love, one that permeates the whole cosmic order, is akin with the formal rational conception of love in Stoic and neo-Stoic philosophy as we examined in Chaps. 3 and 4, with the exception that the Stoic notion of love is grounded solely in the natural world and not in another separate external dimension as in the case of Plato's two-world system that is more akin to Christian cosmology.[4] What is significant about that, however, is that insofar as AGI agents, including superintelligent AI agents (henceforth, I shall refer to those terms synonymously) are rational autonomous agents capable of rational judgements, choices and actions, they would also be capable of loving in the higher form of rational and universal love that Plato and similarly the Stoics envisage. Second, Plato's gradated notion of formal universal love, is similar to the Stoic and neo-Stoic notion of *oikeiosis* we examined earlier in this chapter and in more detail in Chap. 3. That is, the progressive ascent through wisdom and virtue from self-preservation to acquiring a universal normative and cosmic perspective that is inclusive of all humanity, and all rational beings that will also include AI agents. As we saw earlier in this chapter, in the transition from developmental to constitutive virtue, from imperfect and incomplete virtue to perfect and complete virtue, requires a kind of ethical *transformation.*

It is the realisation of the practical necessity for such transformation and the need for providing a conceptual account of how such a transformation can practically take place, that the Stoics developed their theory of oikeiosis. As we saw in Chap. 3 and earlier in this chapter, the four stages comprising the central developmental stages in the Stoic doctrine of oikeiosis

[4] For a discussion of Plato's cosmology, see his dialogue *Timaeus.* 10. Interestingly, Kate Darling in her new book (April 2021) *The New Breed: What Our History With Animals Reveals About Our Future With Machines,* explores our future relationship with machines to our historical relationship with animals. Though her focus is with ordinary suboptimal, less than human intelligent robots, her research might prove useful in shedding more light on our future relationships with AGI agents.

are: the impulse for self-preservation; the impulse for sociability; and the recognition that one should not merely act appropriately towards oneself and others on the basis of "natural impulses" alone, but more importantly should do so on the basis of their rationality. And finally, the disposition and the practice of always acting according to one's perfected reason and virtue in agreement with nature, leads to the attainment of eudaimonia, considered by Stoic and neo-Stoic philosophy to be entirely within our control. Given that at least in principle a Platonic and similarly a Stoic conception of universal love supported also conceptually through the Stoics' notion of oikeiosis does not exclude AGI agents then the notion of universal love for the common good of all rational beings can also be extended to include AGI agents. For AGI agents being rational and capable of being motivated by a similar higher universal love would at least conceptually and rationally be expected not to cause us as harm but moreover, to also promote our good, which is also consistent with the DOIT-Wisdom model we examined in Chap. 2. Therefore, it follows, that both humans and AGI agents have a normative rational motivation at once, epistemic, ethical and eudaimonic, not to cause each other harm but insofar it's within their capacity, to promote each other's good.

In her review of Martha Nussbaum's *Cultivating Humanity: A Classical Defense of Reform in Liberal Education* (1997), Marilyn Friedman (2000) points out that the "idea of love for all humanity," is mentioned by Nussbaum "no less than fourteen times" (Nussbaum 1997). Nussbaum's "love for all humanity" seems essentially similar to the "love of the Good" which, as the primary motivation for doing what is right, as in the case of Plato, is also present in Stoics and neo-Stoic philosophy.

The practical question of course that follows this conclusion is whether the concept of universal rational love can be designed or at least enculturated into AI agents as one of our important human objectives in life, that is, to love and be loved. As Russell, informs us with regard to his beneficial solution to the goal-alignment problem, "machines will need to learn more about what we really want from observations of the choices we make and how we make them" (2019). And aren't those wants, and choices for most people worldwide, have to do, more often than not, with what and who we love in our lives?

From the above, admittedly speculative discussion, insofar as it is techno logically feasible, then there doesn't seem to be a conceptual and theoretical normative reason why the notion of love, given its human central importance and ubiquity, throughout human civilisation, cannot be designed into our objectives the AI agents will learn about us, just as all the other normative principles we examined earlier in this chapter such as equality, solidarity, justice, autonomy and dignity.

As Philip Larkin puts it in the last line of his poem An Arundel Tomb, "What will survive of us is love." If we know that love exists and is experienced between humans and animals such as dogs, cats, parrots, rabbits and guinea pigs, to mention but a few, notwithstanding the vast intellectual disparity and intelligence between us and animals, then analogously it is conceptually feasible that love as an important aspect of human lives, could be developed and be experienced, between us and AGI machines in the form of say, embodied robots, such as Data, in *Star Trek*, and the Hologram Doctor in *Voyager*, for example. Conceptually at least there doesn't appear to be any rational reason why that is not feasible, though there might be technical complications and as yet unforeseen social reasons. As Tina Turner tells us in her song, "its only logical." Time will tell. What we do know, however, is that love and affection is species-neutral and transcends species as our own relationship with animals demonstrates.

What is required, however, as discussed earlier in this chapter, is no less than a transformation to become the best we can be as human beings. For as the Stoics and neo-Stoics show we have the capacity and potential as rational beings imbued with the same cosmic rationality that permeates the whole of nature for becoming the best of ourselves through the inculcation of virtues for the attainment of eudaimonia, for the common good of all sentient beings including AI agents. Through such a transformation, our relationship with AI agents can become a symbiotic beneficial relationship, as suggested by Russell, that benefits the whole of nature.

Whereas AGI agents can offer us more intelligence and knowledge to enhance ours, we, in turn, can offer them something more valuable. We can offer them wisdom. We can do that through our own experience over thousands of years of civilisation and evolution of how best to use and apply superintelligence for the overall eudaimonic benefit of us and AGI agents, and collectively for the whole of nature. Such a singularity[5] would indeed be a welcome one. We certainly have the capacity. The question though remains, do we have the willingness? As Gasset profoundly suggests, this question concerns "the struggle of man with his soul"

[5] For an interesting and absorbing book on the concept of singularity see Kurzweill, Ray (2005). London: Penguin Books.

(1941) just as it did with Faust. The beneficial symbiotic relationship suggested above between us and AGI agents by Russell need not require us to sell our soul for more knowledge and intelligence if we act wisely and design the right values in AGI agents, including love, to be not only intelligent but more importantly to be wise.

Ian McEwan in his book Machines Like Me and People Like You (McEwan 2019) seems to suggest as much, in a conversation between the human Miranda and the AGI Robot, Adam. To her comment that Adam looks sad, he responds thus:

> A self, created out of mathematics, engineering, material science and all the rest. Out of nowhere. No history – not that I'd want a false one. Nothing before me. Self-aware existence. I'm lucky to have it, but there are times when I think that I ought to know better what to do with it. What it's for. Sometimes it seems entirely pointless.

Though highly intelligent, at least as intelligent as Miranda, Adam does not have the skill nor capacity of exploring his existential angst in the way that a human wise person might do. For wisdom, as we explored in earlier chapters is experiential knowledge, a meta-knowledge of how to manage one's first order knowledge of the world. If AGI machines can be designed to develop rational minds that that though far more intelligent, but not as wise as us humans, is there any rational reason, barring any unforeseen technological difficulties, why rational true minds, those of humans and AI machines, cannot love and care for each other's eudaimonic good as conceived by Plato and the Stoics, through the overarching concept of a universal love that permeates, as the Stoics? A Love that is imbued with Wisdom, for as Diotima tells us, "Eros who is in love with what is Beautiful, is a lover of Wisdom" (Plato 1997).

Love Is All Around Us

In the final section of this chapter, I explore some theories or views about the Philosophy of Love (*Oxford Handbook of the Philosophy of Love* (2017). https://doi.org/10.1093/oxfordhb/9780199395729.001.0001

According to Susan Wolf, in her Chapter *"Love: The Basic Questions"*:

> Among the Big Questions that philosophers (and others!) ask, none are older (or, arguably, bogger) than questions about how to live. What are the

> ingredients of a good human life? Is there a single thing that is most important, that any good life must include? …Among the multitude of candidates proposed as essential to good life, one stand out for me as by far the most plausible: love. And of course, it is proposed a lot: Songs, novels, movies, not to mention framed needlepoint pictures, proclaim that "Love makes the world go round"; "Love conquers all"; "All we need is love"; "Tis better to have loved and lost than never to have loved at all."

She goes on to say more specifically with regard to those we love, that "love roots us motivationally to the world in a way that self-interest fails to do…it is a the only possible place, in which you and your loved one can meet – it is the plane of existence you share with the object of your love." Which explains the tragedies of Shakespeare's *Romeo and Juliet* and Richard Wagner's *Tristan and Isolde,* and in Greek Mythology that of Orpheus and Euridice, and the real-life tragedy of the star-crossed lovers, Abelard and Heloise, to name but a few.

Susan Wolf, Chapter 1, Love: The Basic Question, *The Oxford Handbook of the Philosophy of Love* (2017).

In *Iris Murdock on Love,* Niklas Forsberg tells us that in her work "On 'God' and 'Good,'" she claims, "We need a moral philosophy in which the concept of love, so rarely mentioned now by philosophers, can once again be made central." Iris Murdoch, "On 'God' and 'Good,'" in Existentialists and Mystics: Writings on Philosophy and Literature, ed. Peter Conradi, foreword by George Steiner (New York: Penguin, 1999[1997]), 337: "She is asking us to pay attention to love and to think about why it has been blotted out, and by which means." The kind of blockings, according to Forsberg, are due to the fact that the words and things and thoughts that we live by are so ordinary, familiar and homely, that we do not think about them and perhaps also because we do not think about them perhaps because we do not expect to find philosophy there. As he correctly observes this is well illustrated by Wittgenstein's famous remark:

> The aspects of things that are most important for us are hidden because of their simplicity and familiarity. (one is unable to notice something – because it is always before one's eyes.) The real foundations of his enquiry do not strike a man at all. Unless *that* fact has at some time struck him. – And this means we fail to be struck by that, once seen, is most striking and powerful. (Ludwick Wittgenstein, *Philosophical Investigations*, 2nd ed., trans. G.E. M. Anscombe (Malden, MA: Blackwell, 1997)

According to Forsberg, "love has struck Murdoch – informed by reading Plato's works and love's absence in philosophy is to her most striking and powerful. So, by this retracing of concerns, we get to the beginning: Plato. For the inheritance of Plato comes into view in her emphasis on *Eros* but also in her claim that philosophy is, or should at least be, 'anamnesis', 'recollection' of what we did not know we knew." Love's importance, the force that Eros signifies, is for Murdoch indisputable, and this makes love's absence I philosophy a central problem. As we saw earlier Plato's Symposium, on, Eros is not the Good, he is not even a god, but a spirit which moves towards the good.

According to Murdoch,

> We need a moral philosophy in which the concept of love, so rarely mentioned by, philosophers. Can once again be made central. (Murdoch, "On 'God' and 'Good'")

In *Aristotle on Love of Friends*, Neera K, Badhwar and Russel E. Jones, tells us that in the *Nichomachean Ethics* devote 20 percent to a discussion of friendship for as they state "No one would choose to live without friends.... not even if he has every other good thing." According to these authors, Aristotle argues that "friends show their trustworthiness and loyalty to each other by forgiving each other's forgivable faults and help each other to become better. They show their trust in each other by being open to correction. They therefore show virtue in response to their own and each other's imperfections. Their relationship is therefore grounded both in what is already admirable in each other. By delighting in the best of each other, both can also see and more fully actualise the best in themselves." In Aristotle, as in the Stoics that follow, love and friendship are anchored in the virtues of prudence, fortitude, justice temperance and moderation, leading to the attainment of eudaimonia, well-being or happiness. Both Aristotle and later the Stoics indicating the importance of a life lived virtuously for the attainment of eudaimonia. A life that regrettably Aristotle's student Alexander the Great failed to accomplish.

The philosophy of Baruch Spinoza (1632–1677) is of particular importance to the philosophy of love as the conception of love lies at the heart of his ethics as Michael Strawser argues in his book, *Spinoza and the Philosophy of Love*, 2021.

According to Strawser,

> Spinoza clearly connects the knowledge of God with love. Spinoza suggest that knowing God amounts to performing deeds of love. Spinoza, he argues, directly equates Gor with Nature. For Spinoza it is no doubt important to grasp the nature of this unified whole – *Deus sive Natura* (God, or in other words Nature) – in order to understand the nature of the parts. This equivalence ….is significant when we consider love as an active emotion manifesting the power of human being, as well as when we consider the "love of God," understood in both the directions from God to human beings and from human beings to God. Spinoza according to Strawser, "clearly connects the knowledge of God with love…. A love that ultimately advises us to help our neighbours by desiring the good that we desire for ourselves for all other persons as well – and the more we do this, Spinoza explains, the more we have a 'greater knowledge of God.'" …. For being led by love is leading by love, which amounts to our greatest good or supreme happiness. Further, Spinoza indicates here that happiness itself is (also) "virtue and the service of God….and perfect freedom." We can conceive, Strawser tells is, "of an expression of love qualified by virtuous activity, in other words, virtuous love."

Having examined the characteristics of Stoic philosophy in this book and particularly in Chap. 3, we can clearly see now some striking similarities between Spinoza's and Stoic philosophy, especially the connection with God and Nature, and a virtuous life leading to happiness for us and for our all our fellow human beings, worldwide. For the Stoics identified just as Spinoza, Nature with God and Nature/God with Love.

The following poem, A Perfect Boy, I wrote that was selected and published in 2023 in the *Australian Catholic University* (ACU),[6] based on a real experience I witnessed in Larnaca, Cyprus about Ilya and his mother. Ilia, a young boy, suffers from cerebral palsy and is in a wheelchair and can't talk. The poem expresses the love of his mother for the boy, as I witnessed it.

A Perfect Boy

Dedicated to Ilya

He sat in a wheel-chair in the shape of a pram
His head flopping to one side

[6] Love. Poems from the 2023 ACU Prize for Poetry. ISBN:978-1-922325-21-1

As his mother wheeled the pram along the promenade, always
This time of day, So Ilya
Could throw pebbles she'd picked for him
Some white, some black, loping them one by one
over the boardwalk, into the Mediterranean sea
And when the bucket in which she placed the pebbles
Was half-full or half-empty
Ilya would toss it back into the sea
As his mother climbing barefoot down the crashing waves
Filled the bucket once again with black and white pebbles
Then wiping the saliva flowing like lava
Down the side of his twisted mouth, staining
His new clean clothes, his mother had pressed for him that morning,
With excitement or frustration, or some deep, hidden sorrow
Ilya would throw the bucket of pebbles untouched back into the sea
As the swallows sensing an open day for play, swooped
And dived and curved and turned vertical and horizontal,
This way and that way, over
The child's head, chirping, while, he screamed with laughter
Trying to clap with one hand, Clawed and bent
As once he saw a seal do the same, but instead
Stamped his feet in released delight
And watched the swallows flying away like fighter planes over the sea
Throwing the remaining pebbles to the playful waves
While his mother smiled with pride knowing
Her son, was nothing less,
Than a perfect boy,
Since he loved swallows
As much as she loved him –
And leaning over,
She wiped the sea-spray mixed with his saliva
From his gentle, troubled face.

(Edward H. Spence, A Perfect Boy, *Love,* 2023 ACU Prize for Poetry, 170–171)

If AGI (Artificial General Intelligent) machines can be designed to develop rational minds that though far more intelligent, but not as wise as us, is there any rational reason, barring any unforeseen technological difficulties, why rational true minds, those of humans and AI machines, cannot love and care for each other's eudaimonic good through the overarching concept of a universal love as conceived by Plato and the Stoics? I conclude, with Sonnet 116 by William Shakespeare who also conceived of such a universal love that permeates, as the Stoics believed, the whole of nature. A Love that is imbued with Wisdom, for as Diotima tells us, "Eros who is in love with what is beautiful, is a lover of wisdom" (Plato 1997).

I conclude this final section of the chapter with a sonnet by a very wise poet, who, like Plato before him, recognised that love is ultimately a love between true minds.

> Let me not to the marriage of true minds
> Admit impediments. Love is not love
> Which alters when it alteration finds,
> Or bends with the remover to remove.
> no! it is an ever-fixed mark
> That looks on tempests and is never shaken;
> It is the star to every wand'ring bark,
> Whose worth's unknown, although his height be taken.
> Love's not Time's fool, though rosy lips and cheeks
> Within his bending sickle's compass come;
> Love alters not with his brief hours and weeks,
> But bears it out even to the edge of doom.
> If this be error and upon me prov'd,
> I never writ, nor no man ever lov'd.
> *Shakespeare, Sonnet 116*

References

Camus, Albert. (2000). *The Myth of Sisyphus.* Translated from the French by Justin O'Brien. Harmondsworth, Middlesex: Penguin Modern Classics.

D'Angour, Armand. (2019). *Socrates in love: The making of a philosopher.* Bloomsbury Publishing.

Friedman, M. (2000). "Educating for World Citizenship." *Ethics, 110*(3), 586–601. https://doi.org/10.1086/233325

Gasset, Jose Ortega Y. (1941). "Man, the Technician," Chapter 3. *Toward A Philosophy Of History.* New York: W. W. Norton.

Grau, Christopher and Smuts, Aaron. (2017). *Oxford Handbook of the Philosophy of Love.* Oxford: Oxford University Press.

Kurzweill, Ray. (2005). *The singularity is near: when humans transcend biology.* Penguin Books: New York.

Spence, Edward H. (2021) *Stoic Philosophy and the Control Problem of AI Technology: Caught in the Web*, Rowman & Littlefield Publishers, Maryland, USA.

McEwan, Ian. (2019). *Machines Like Me and People Like You.* London: Jonathan Cape.

Nussbaum, Martha. (1997). *Cultivating Humanity: A Classical Defence of Reform in Liberal Education.* Cambridge, Massachusetts: Harvard University Press:

Plato. (1997). *Plato: Symposium and Death of Socrates.* Translated by Tom Griffith. Hertfordshire: Wordsworth Classics.

Russell, Stuart. (2019). *Human Compatible: AI and the Problem of Control.* UK: Penguin Books.

Tegmark, Max. (2018). Life 3.0: *Bring Human in the age of Artificial Intelligence.* UK: Penguin Books.

CHAPTER 9

Epilogue

Great lords, wise men ne'er sit and wail their loss,
But cheerly seek how to redress their harms.
—Shakespeare (Henry VI, Part 3 [act V, Scene 4])

Some Future Speculations

This book concludes by way of reasoned speculation about some theoretical and practical problems that may potentially arise through our future relationship with AI machines and their impact on our well-being. Could wisdom be humanity's secret weapon in ensuring that AI machines far more intelligent than humans live peacefully with us and do not enslave us nor seek to destroy us? Moreover, could we be, as James Lovelock speculates in his last book *Novacene* (2020), "all watched over by machines of Loving Grace"?

An important and central question that has arisen prominently recently concerning AI among academics, experts and generally the media, is whether AI can be conscious. "Artificial intelligence systems capable of feelings or self-awareness are at risk of being harmed if the technology is developed irresponsibly," according to an open letter signed by AI practitioners and thinkers" reports Dan Milmo, Global Technology editor. Citing a paper written by Oxford University's Patrick Butlin and Theodoros Lappas of the Athens University of Economics and Business, he goes on

E. H. Spence, *Wisdom in the Age of Intelligent Machines*,
https://doi.org/10.1007/978-3-031-93883-2_9

to say "that, it may be the case that large numbers of conscious systems could be created and caused to suffer,... adding that if powerful AI systems were able to reproduce themselves it could lead to the creation of large numbers of new beings deserving moral consideration." He goes on to add "that even companies not intending to create conscious systems will need guidelines in case of 'inadvertently' creating conscious entities and acknowledges that there is widespread uncertainty and disagreement over defining consciousness in AI systems and whether it is even possible, but says it is an issue that 'we must not ignore.'" Citing Sir Demis Hassabis, the Head of Google's AI programme and a Nobel Prize winner, Dan Milmo reports that according to Hassabis, AI systems were "definitely" not sentient currently but could be in the future. "Philosophers haven't really settled on a definition of consciousness yet but if we mean sort of self-awareness, these kinds of things, I think there's a possibility AI one day could be" he said in an interview with US broadcaster CBS, Tue 4 Feb 2025.

In the journal paper cited above by Dan Milmo, "Principles for Responsible AI Consciousness Research" by Patrick Butlin, Global Priorities Institute, University of Oxford, and Theodoros Lappas Department of Marketing & Communication Athens University of Economics and Business, Athens 10434, Greece, the authors claim

> that organisations pursuing AI consciousness research—or advanced AI research more broadly—should adopt principles to mitigate the risks outlined in the previous section.

The risks referred by the two authors are:

First, where people believe that AI systems or AI-generated characters are conscious, this could increase their use of these systems for companionship and interaction, and deepen the emotional bonds they feel in relation to them. The relationships people have with AI systems or characters could disrupt valuable human relationships, although they could also be valuable in themselves. The use of AI for social functions is likely to increase anyway, but it may have more powerful effects if AI systems give a more compelling impression of consciousness.

Second, belief in AI consciousness could increase trust in the systems concerned, and thus reliance on them, including willingness to follow suggestions and disclose information. There is evidence that anthropomorphism and a feeling of "closeness" between users and AI systems lead to increased trust (Bach et al. 2022). This could have either harmful or beneficial effects, depending on whether the systems concerned are in fact trustworthy.

Third, belief in AI consciousness could lead to calls for AI rights and greatly intensified debate on the topic among experts and the public. While we believe that informed discussion of AI consciousness is necessary, some of the consequences of movements for AI rights could be problematic. Misguided efforts to promote the welfare of AI systems that are not in fact moral patients could lead to misallocation of resources, of concern, or of political energy. Such efforts could be wasteful or harmful, and could slow innovation and deployment, leading to a significant loss of potential benefits from AI. consciousness become entrenched it could take decades for these conditions to ease. These considerations all illustrate the importance of ensuring that public discussion of AI consciousness is well-informed, as far as possible, from the outset.

The intention of the five principles suggested by the two authors as stated below is that by adopting these principles, or similar ones, organisations will be "more likely to achieve two higher-level goals. These are: to avoid contributing to any future mistreatment of AI moral patients; and to promote understanding of concepts, arguments and evidence concerning consciousness among the public and professionals in relevant fields. In this section, we explain our proposed principles and discuss some related issues. The five subsections of this section correspond to our five principles:

(1) Objectives: Organisations should prioritise research on understanding and assessing AI consciousness with the objectives of (i) preventing the mistreatment and suffering of conscious AI systems and (ii) understanding the benefits and risks associated with consciousness in AI systems with different capacities and functions.
(2) Development: Organisations should pursue the development of conscious AI systems only if doing so will contribute significantly to the objectives stated in Principle 1, and (ii) effective mechanisms are employed to minimise the risk of these systems experiencing and causing suffering.
(3) Phased approach: Organisations should pursue a phased development approach, progressing gradually towards systems that are more likely to be conscious or are expected to undergo richer conscious experiences. Throughout this process, organisations should (i) implement strict and transparent risk and safety protocols and

(ii) consult with external experts to understand the implications of their progress and decide whether and how to proceed further.

(4) Knowledge sharing: Organisations should have a transparent knowledge sharing protocol that requires them to (i) make information available to the public, the research community and authorities, but only insofar as this is compatible with (ii) preventing irresponsible actors from acquiring information that could enable them to create and deploy conscious AI systems that might be mistreated or cause harm.

(5) Communication: Organisations should refrain from making overconfident or misleading statements regarding their ability to understand and create conscious AI. They should acknowledge the inherent uncertainties in their work, recognise the risk of mistreating AI moral patients, and be aware of the potential impact that communication about AI consciousness can have on public perception and policy making."

Another important and in my opinion, essential paper published by a group of notable academics recently on AI consciousness is that of ***Talking AI Welfare Seriously,*** (arXiv:2411.00986v1 [cs.CY] 4 Nov 2024) as these authors state and I include below for completion of the actual statement:

> In this report, we argue that there is a realistic possibility that some AI systems will be conscious and/or robustly agentic in the near future. That means that the prospect of AI welfare and moral patienthood — of AI systems with their own interests and moral significance — is no longer an issue only for sci-fi or the distant future. It is an issue for the near future, and AI companies and other actors have a responsibility to start taking it seriously. We also recommend three early steps that AI companies and other actors can take: They can (1) acknowledge that AI welfare is an important and difficult issue (and ensure that language model outputs do the same), (2) start assessing AI systems for evidence of consciousness and robust agency, and (3) prepare policies and procedures for treating AI systems with an appropriate level of moral concern. To be clear, our argument in this report is not that AI systems definitely are — or will be — conscious, robustly agentic, or otherwise morally significant. Instead, our argument is that there is substantial uncertainty about these possibilities, and so we need to improve our understanding of AI welfare and our ability to make wise decisions about this issue. Otherwise, there is a significant risk that we will mishandle decisions about AI welfare, mistakenly harming AI systems that matter morally and/or mistakenly caring for AI systems that do not.

Robert Long, Eleos AI.; Patrick Butlin, University of Oxford; Jacqueline Harding, Stanford University; Jonathan Birch, London School of Economics; Jeff Sebo, New York University; Kathleen Finlinson, Eleos AI; Jacob Pfau, New York University; Kyle Fish, Eleos AI, Anthropic; Toni Sims, New York University; and David Chalmers, New York University.

In conclusion of their report the authors conclude, and state below for completion,

We have argued that there is a realistic, non-negligible chance that some AI systems will be welfare subjects and moral patients in the near future, given current evidence. When we consider all relevant issues in philosophy, science and technology with sufficient care, it becomes difficult to simply dismiss the idea of near-future AI moral considerability out of hand. That would require having a very high degree of confidence in a very restrictive set of views about some of the hardest problems in philosophy, science and technology, ranging from the nature of moral patienthood to the nature of consciousness to the future of AI. We are simply not warranted in consistently having this much confidence in these kinds of views at this stage.

To be sure, these reflections are far from conclusive. In the long run, there is no substitute for rigorous, systematic, integrative assessment of all the issues that we discuss here. We need to develop comprehensive, or at least representative, lists of possible bases of welfare and moral patienthood; of possible bases of consciousness, robust agency and other potentially morally significant capacities; and of reliable indicators for all of these features in AI systems. We also need to continue to do research in the humanities, social sciences and natural sciences so that we can gradually improve our distribution of credence across these possibilities over time. Indeed, the need for this research is one of our main conclusions here. But these reflections are still suggestive. AI development is proceeding at a rapid pace, and technological change tends to be faster than social change. That means that we need to think ahead and do the best we can with what we have. And for as long as detailed tools for assessing AI moral patienthood remain unavailable, these reflections can at least serve as a useful corrective to the assumption — which might otherwise seem plausible to some — that AI welfare and moral patienthood is either not a problem at all or a problem only for the distant future. When we think about how many bold assumptions are required for this assumption to be secure, we should accept that further assessment and reasonable precautions are warranted.

In this report, we presented three general steps that leading AI companies can and should take immediately regarding AI welfare. First, they should **acknowledge** that AI welfare is a legitimate issue that merits attention now, and they should communicate about this issue with caution and humility. Second, they should assess leading AI systems for evidence of welfare and moral patienthood, by examining these systems for architectural markers of consciousness and robust agency. Third, they should prepare policies and procedures that will allow them to extend an appropriate level of moral concern to AI systems in the near future, by hiring or appointing an AI welfare officer and building thoughtful guidelines with expert and public input.

As we have now repeatedly emphasised, taking these first steps regarding AI welfare will not be enough. It will take time for researchers, companies, governments and other leaders in this space to develop a shared infrastructure for adequately assessing and addressing this issue. However, by taking these first steps, leading AI companies can play an essential role in making this further work possible. Given that leading AI systems already possess some markers of consciousness and robust agency, given that further AI development might be quite fast, and given that the development of an adequate policy response might be quite slow, leading AI companies should take these steps as soon as possible.

Our team is conducting further research on AI welfare and moral patienthood, with special focus on developing a research agenda towards (a) assessing AI systems for consciousness and robust agency and (b) preparing policies and procedures for extending an appropriate level of moral concern to AI systems. In the meantime, we urge AI companies to take the first steps described here."

The above arguments and statements in the two papers as summarised above, indicate, I hope clearly, that the topic of AI consciousness is now taken quite seriously, by a number of researchers and experts on AI, across the board. In his final book *Novacene*, James Lovelock, with whom I began this final section of this book, says something quite interesting and prophetic about our relationship with AI agents, which he refers to as *cyborgs*:

> If I am right about the Gaia hypothesis and the Earth is indeed a self-regulating system, then the continued survival of our species will depend on the acceptance of Gaia by the cyborgs. In their own interest, they will be obliged to join us in the project to keep the planet cool. .... This is why I

> believe the idea of a war between humans and machines or simply the extermination of us by them is highly unlikely. Not because of our imposed rules, but because of their own self-interest, they will be eager to maintain our species as collaborators. (Lovelock 2020)

For this reason, at least, it would be easier to communicate with AI agents if they were conscious who could better understand and hopefully appreciate who we are. Then perhaps this is why wisdom is as important for us as it is for AI agents, and as I have argued previously in Chap. 4 in this book, regarding my notion of a hybrid notion of AI Wisdom, our ongoing relationship with AI agents, is a matter of "collaboration" or as James Lovelock speculates in his last book *Novacene* (2020), "all watched over by machines of Loving Grace."

As Susan Schneider points in her enlightening book, *Artificial You: AI and the Future of Your Mind* (2019),

> From an ethical standpoint, it is best to assume that a sophisticated AI may be conscious, at least until we develop tests, for consciousness that we can have confidence in….Not only could failure to recognize a machine as sentient cause needless pain and suffering, but, as films like ex Machina and I, Robot illustrate, any failure to be charitable to AI may come back to haunt us, as they may treat us as we treated them….a cautious approach is most prudent.

Schneider goes on to wisely conclude, as we would expect Socrates to do,

> As we consider mind-design decisions, we must do so, first and foremost with a mindset of metaphysical humility. Remember the risks. The future of the mind, whether the minds are human minds or robot minds, is a matter that requires public dialogue and philosophical contemplation. (Schneider, Susan. 2019)

And David Robson I think will concur that "intellectual humility and actively open-minded thinking – can help us to navigate our way around the intelligence trap" (David Robson (2019), Hodder and Stoughton, Hachette: London, UK (113)).

As Justine E. H. Smith quite sensibly and correctly points out, "in order to justify the supposition that AI could eventually become conscious, we would have to be confident in our understanding of what consciousness is. And there is simply nothing about the present state of our knowledge, empirical or theoretical, to justify such confidence, Consciousness very well *might* be something exhaustively explicable in terms or algorithmic processes that may be instantiated indifferently in carbon-based, or silicon-based, or for that matter a string-and toilet-paper-roll-based material substratum. But no one has yet proven that it is, or even made any significant progress towards showing how such a thing *might* be proven. As David Chalmers correctly pointed out, in his 1996 book *The Conscious Mind: In Search of a Fundamental Theory*, "consciousness is a really hard problem."

Thinking Weapons

James Lovelock in his last book *Novacene* (2020) referred to earlier in this chapter, talks of AI weapons (Chapter 21). Referring to the V-I missiles dropped on London by the Germans in the Second World War, Lovelock refers to a comment made by a civilian who exclaimed: "What on earth is happening?" (112). Lovelock rightly expresses his concern thus: "The notion of allowing the evolution of adaptive computer systems on military platforms seems to me to be potentially the deadliest idea yet introduced for the replacement of human and other organic life on Earth. By taking this route, we would set in progress the evolution of a new life form that would emerge as a soldier equipped with the latest and most deadly weapons."

Ironically, as Lovelock informs us "in July 2017 Elon Musk and 115 other Silicon Valley AI specialists wrote an open letter to the UN, asking a ban on autonomous weapons … known in the trade as LAWS – lethal autonomous weapons systems – these are devices that can seek, identify and kill enemy targets."

Tragically, those AI lethal autonomous weapons were recently used by Israel in Gaza against the Palestinians to indiscriminately, not only to kill army personnel but non-combatant civilians, including women and children. The International Court of Justice (ICJ) has examined and found there are grounds for the charge of genocide brought recently by South Africa against Israel at the UN.

Ironically, Elon Musk, who is now a member of the US government under the Presidency of Donald Trump, has been busy using AI to remove the checks and balances previously in place by the various civil government departments and institutions in the US. Trump himself has expressed an opinion that the US should take over the Palestinian lands in Gaza and turn them into a new "Riviera." This of course is only empty speculation and hopefully the Palestinians and the Israelis would come to a mutual peaceful agreement amicably and ethically, from both sides. Time will tell.

In the meantime, however, since Trump's implementation of his "Liberation Day" in early April 2025, involving the imposition of worldwide "tariffs" has plunged the world economies, including the EU, as well as the USA itself, in an economic meltdown across the globe, with political implications, potentially affecting world peace. Contrary to the Stoic notion of a cosmopolitan world based on ethics and the cardinal virtues of justice, moderation, courage and prudence, for the good and well-being of everyone, guided by the principles of Love and Wisdom, as we have examined in this book.

Also ironically, the current AI announcement from the *AI Summit in Paris The International Association for Safe and Ethical Artificial Intelligence (IASEAI)* (2025) also seems to put on notice Elon Musk and the other Big Tech companies in the Silicon Valley currently and President Trump's activities as their current ambition to undermine the EU, who have issued a call to action:

"For Lawmakers, Academics, and the Public Ahead of AI Summit in Paris," the International Association for Safe and Ethical Artificial Intelligence (IASEAI) concluded its inaugural conference today by calling for ten critical action items. Feb. 7, PARIS: The IASEAI '25 conference has been the largest ever global gathering focusing on cutting-edge developments in AI safety and ethics. Held at the OECD headquarters just prior to the Paris AI Action Summit, the conference brought together leading representatives from academia, civil society, industry, media and government. Professor Stuart Russell, IASEAI President and Distinguished Professor of Computer Science at the University of California, Berkeley said: "The development of highly capable AI is likely to be the biggest event in human history. The world must act decisively to ensure it is not the last event in human history. This conference, and the cooperative spirit of the AI Summit series, gives me hope; but we must turn hope into action, soon, if there is to be a future we would want our children to live in." The following actions, derived from the research presented and the deliberations held during the conference, will help to further IASEAI's mission to "ensure that AI systems are guaranteed to operate safely and ethically." IASEAI calls for:

1. Recognition of the significance of new developments in AI. Policymakers must act with an urgency that matches the transformational potential of AI, the rapidity of change, and the increasing risks to humanity as AI capabilities begin to exceed our own.
2. Preventing AI-driven institutional and social disruption. The power of AI threatens to disrupt employment and social structures, worsen inequality and severely compromise the information ecosystem. Policymakers must take proactive steps to protect institutions, individuals and ways of life while harnessing AI to strengthen rather than weaken societies. To the extent that these problems take similar forms across nations, policymakers should collaborate to seek collective solutions.
3. Addressing the race to AGI. The financial and strategic incentives to achieve artificial general intelligence (AGI) or "superintelligence" lead companies and countries to undercut safety standards in an attempt to gain technological and political control. Policymakers should take coordinated actions in international fora such as the UN and the OECD to ensure that innovation and competition proceed within an agreed framework of rigorous safety standards.
4. Coalescing the efforts of research communities around the goal of safe and ethical AI. The increasing threats to human flourishing posed by AI systems require researchers from the AI, ethics, social science and policy communities to collaborate and pool their efforts. The various perspectives offered by these communities are not in tension; on the contrary, they all contribute in important ways to ensuring that AI systems do not harm human society.
5. Adoption of mandatory safety and ethical requirements. While commendable, voluntary commitments by companies must be made more specific and legally binding. Such binding commitments might include mandatory registration of advanced AI systems, including automatic self-registration for open-source copies; installation of remotely operable off-switches; mandatory reporting of incidents; professional ethics training for engineers; and standards for training, design, development, testing, auditing and transparency of advanced AI systems. Developers should provide a scientifically convincing safety case that their systems will not cross so-called behavioural red lines, which demarcate unacceptable behaviours.

6. Advancement of global cooperation. AI developments and the accompanying risks have cross-border impacts, necessitating global cooperation on AI safety research and regulation, considering the perspectives of all nations. Moreover, the benefits of AI must be equitably distributed. In this vein, we welcome the establishment of the UN AI Advisory Body and the international network of AI Safety Institutes.
7. An increase in publicly funded research. The scale of the challenge requires significantly more publicly funded AI safety and ethics research. Research cannot continue to be dominated by companies with significant conflicts of interest. The potentially transformative benefits of AI can be realised only if advances in capabilities are accompanied by methods to ensure that AI systems are safe by design and aligned with human interests.
8. Support of the AI Foundation. The forthcoming AI Foundation, as anticipated by the Paris Summit hosts, seems to be a significant step forward. The Foundation must support the design and development of AI systems to address human needs and enhance local capabilities, and governments must, in turn, support the Foundation.
9. Support of the Council of Europe Framework Convention on Artificial Intelligence. The Council's AI Treaty "aims to ensure that activities within the lifecycle of artificial intelligence systems are fully consistent with human rights, democracy and the rule of law." While the Council administers the treaty, it is open for signature by non-member states. A total of 37 countries have signed, including the United States and United Kingdom, and more are expected to sign at the Paris Summit.
10. Fostering informed dialogue. AI researchers and the media need to contribute their complementary forms of expertise in informing the public and policymakers, objectively and accurately, about developments in AI and their significance. For more information about IASEAI, the conference, the actions, or to speak with those involved, please contact Jess Graham (jessgraham@iaseai.org). About IASEAI: The International Association for Safe and Ethical AI (IASEAI) is a non-profit organisation founded to address the risks and opportunities associated with rapid advances in AI. Their mission is to ensure that AI systems are guaranteed to operate safely and ethically; and to shape policy, promote research, and build understanding and community around this goal."

I conclude with some profound words from Lovelock, whose Gaia, like the Earth of the Stoics, they considered divine and of which we are part and therefore ought to be respectful, considerate, virtuous and ethical towards all other humans, and all life on Earth, which the Stoics considered all humans, as our fellow cosmopolitans:

That which we are, we are. That is the wisdom of great age, the acceptance of our impermanence while drawing consolation from the memories of what we did and what, with luck, we might yet do. Also perhaps, we can hope that our contribution will not be entirely forgotten as wisdom and understanding spreads outwards from the Earth to embrace the cosmos.

Ulysses by Alfred, Lord Tennyson
It little profits that an idle king,
By this still hearth, among these barren crags,
Match'd with an aged wife, I mete and dole
Unequal laws unto a savage race,
That hoard, and sleep, and feed, and know not me.
I cannot rest from travel: I will drink
Life to the lees:
……………………………………………
There lies the port; the vessel puffs her sail:
There gloom the dark, broad seas. My mariners,
Souls that have toil'd, and wrought, and thought with me—
That ever with a frolic welcome took
The thunder and the sunshine, and opposed
Free hearts, free foreheads—you and I are old;
Old age hath yet his honour and his toil;
Death closes all: but something ere the end,
Some work of noble note, may yet be done,
Not unbecoming men that strove with Gods.
The lights begin to twinkle from the rocks:
The long day wanes: the slow moon climbs: the deep
Moans round with many voices. Come, my friends,
'T is not too late to seek a newer world.
Push off, and sitting well in order smite
The sounding furrows; for my purpose holds
To sail beyond the sunset, and the baths
Of all the western stars, until I die.
It may be that the gulfs will wash us down:
It may be we shall touch the Happy Isles,

And see the great Achilles, whom we knew.
Tho' much is taken, much abides; and tho'
We are not now that strength which in old days
Moved earth and heaven, that which we are, we are;
One equal temper of heroic hearts,
Made weak by time and fate, but strong in will
To strive, to seek, to find, and not to yield.

References

Bach, T.A., Khan, A., Hallock, H., Beltrao, G., Sousa, S. (2022). "A Systematic Literature of Review of User Trust in AI-Enabled Systems: An HCI Perspective." *International Journal of Human-Computer Interaction*, 40(5):1251–1266.

Chalmers, David J. (1996) *The Conscious Mind - In Search of a Fundamental Theory*. Oxford: Oxford University Press.

International Association for Safe and Ethical AI (IASEAI). Inaugural Conference Feb 6–7, 2025. Paris, France. https://www.iaseai.org/conference

Lovelock, James. (2020) *Novacene - The Coming Age of Hyperintelligence*. Penguin Press.

Robson, David. (2019). *The Intelligence Trap – Revolutionise Your Thinking and Make Wiser Decisions*. Hodder and Stoughton: London, UK.

Schneider, Susan (2019). *Artificial You: AI And The Future of Your Mind*: Princeton and Oxford: Princeton University Press.

Bibliography

Aizenberg, Evgeni and Van Den Hoven, Jeroen. (2020). "Designing for human rights in AI." *Big Data and Society* 1(1): 1–12.

Annas, Julia. (1993). *The Morality of Happiness.* New York: Oxford University Press.

Annas, Julia. (1999). *Platonic Ethics, Old and New.* Ithaca: Cornell University Press.

Annas, Julia. (2004). "Happiness as Achievement." *Daedalus*, 133(2): 44–51.

Annas, Julia. (2011). *Intelligent Virtue.* Oxford: Oxford University Press.

Ardelt, M and Edwards, CA. (2016). "Wisdom a the End of Life: An Analysis of Mediating and Moderating Relations Between Wisdom and Subjective Well-Being." *J Gerontol B Psychol Soc Sci*, 71(3): 503–13

Aristotle. (1941). "Nichomachean Ethics", In Richard McKeon (ed) *The Basic Works of Aristotle*, New York: Random House.

Aurelius, Marcus. (1964). *The Meditations.* Translated by Maxwell Staniforth, Hamondsworth: Penguin Books.

Bach, T.A., Khan, A., Hallock, H. et al. (2022). "A Systematic Literature Review of User Trust in AI-Enabled Systems: An HCI Perspective." *International Journal of Human-Computer Interaction*, 40(5): 1251–1266.

Baltes, PB. and Smith, J. (2004). "Lifespan Psychology: From Developmental Contextualism to Developmental Biocultural Co-constructivism." *Research in Human Development*, 1(3): 123–144.

Baltes, PB., Gluck, J. and Kunzmann, U. (2002). "Wisdom: Its Structure and Function in Regulating Successful Life Span Development." In CR Snyder and SJ Lopez (eds) *Handbook of Positive Psychology*, Oxford: Oxford University Press.

E. H. Spence, *Wisdom in the Age of Intelligent Machines*,
https://doi.org/10.1007/978-3-031-93883-2

Banicki, K. (2009). "The Berlin Wisdom Paradigm: A Conceptual Analysis of a Psychological Approach to Wisdom." *History and Philosophy of Psychology*, 11(2): 25–36.

Becker, Laurence. (1998). *A New Stoicism*, Princeton: Princeton University Press.

Beyleveld, Deryck. (1991). *The Dialectical Necessity of Morality: An Analysis and Defense of Alan Gewirth's Argument to the Principle of Generic Consistency.* Chicago: University of Chicago Press.

Blok, Vincent. (2019). "Towards and Ontology of Innovation: On the New, the Political-Economic Dimension and the Intrinsic Risks in Innovation Process." *Handbook of Philosophy of Engineering*, edited by Diana Michelfelder and Neelke Doorn. Routledge.

Bostrom, Nick. (2015). *Superintelligence; Paths, Dangers, Strategies.* Oxford: Oxford University Press.

Brey, Philip, Briggle, Adam, Spence, Edward H. (eds). (2012). *The Good Life in a Technological Age.* New York: Routledge.

Burr, Christopher, Taddeo, Mariarosaria and Floridi, Luciano. (2020). "The Ethics of Digital Wellbeing: A Thematic Review." *Science and Engineering Ethics* 26: 2313–2343.

Buttazzo, Giorgio. (2001). "Artificial Consciousness: Utopia or Real Possibility." *Computer*, 34: 24–30.

Camus, Albert. (1975). *The Myth of Sisyphus.* Translated from the French by Justin O'Brien. Harmondsworth, Middlesex: Penguin Books.

Cavanna, Andrea E. (2019). "Back to the Future: Stoic Wisdom and Psychotherapy for Neuropsychiatric Conditions." Editorial, *Future Neurology* 14(1), 15 January 2019.

Ceva, Emanuela and Ferretti, Maria. (2017). "Political Corruption." *Philosophy Compass,* 12(2): 1–10.

Chalmers, David J. (1996). *The Conscious Mind - In Search of a Fundamental Theory.* Oxford: Oxford University Press.

Cicero. (1927). *Tusculan Disputations* (Loeb Classical Library) 2nd Edition, Translated by J.E. King. Cambridge, Massachusetts, and London: Harvard University Press.

Davis, Jenny., Williams, Apryl and Yang, Michael. (2021). "Algorithmic Reparation." *Big Data and Society,* 8(2): July.

Descartes, R. (1979). "Principles of Philosophy." In *Philosophical Works of Descartes,* E Haldine and G Ross (Trans. & Eds.), London: Cambridge University Press.

Dretske, Fred. (1999). *Knowledge and the Flow of Information.* Stanford: CSLI Publications.

Dylan, Bob. (1964). *The Times Are A Changin.* bobdylan.com/albums/the-times-are-a-changin/

Elliott, Deni and Spence, Edward H. (2018). *Ethics for a Digital Era.* Oxford: Wiley Blackwell.

Epictetus. (2018). *Epictetus: Discourses and Selected Writings*, translated and edited by Robert Dobbin. London: Penguin Classics.
Erasmus., D. (1994). *In Praise of Folly*. A.H Levi (eds) and B. Radice (trans). UK: Penguin Classics.
Finnis, J. (1980). *Natural Law and Natural Rights*. Oxford: Clarendon Press.
Finnis J. (1983). *Fundamentals of Ethics*. Oxford: Clarendon Press.
Flanagan, Owen. (2009). *The Really Hard Problem: Meaning in a Material Word*. Massachusetts Institute of Technology: MIT Press.
Floridi, Luciano. (2005). "Is Semantic Information Meaningful Data?" *Philosophy and Phenomenological Research*, LXX(2): 351–370.
Floridi, Luciano. (2002). "What is the Philosophy of Information?" *Metaphilosophy*, 23: 123–145.
Floridi, Luciano. (1999). "Information Ethics: On the Theoretical Foundations of Computer Ethics." *Ethics and Information Technology*, 1(1): 37–56.
Floridi, Luciano. (2011). "The Construction of Personal Identities On-line." *Minds and Machines*, 21(4): 477–479.
Floridi, Luciano. (2014). *The Fourth Revolution:How the Infosphere is Reshaping Human Reality*. Oxford England:Oxford University Press.
Friedman, Marilyn. (2000). "Educating for World Citizenship" *Ethics* (April 2000), 589.
Floridi, Luciano. (2013). *The Ethics of Information*. Oxford: Oxford University Press.
Floridi, Luciano. (2019). "The Fight for Digital Sovereignty: What It Is, and Why It Matters, Especially for the EU." *Philosophy and Technology* 33: 369–78.
Foroohar, Rana. (2019). *Don't Be Evil: The Case Against Big Tech*. UK:Allen Lane.
Frischmann, Brett and Salinger, Evan. (2018). *Re-Engineering Humanity*. Cambridge, UK: Cambridge University Press.
Galloway, Scott. (2017). *The Four: The Hidden DNA of Amazon, Apple, Facebook and Google*. London: Bantam Press.
Garrett, R. (1996). "Three Definitions of Wisdom." In Lehrer and Smith (eds), *Knowledge, Teaching and Wisdom*. Dordrecht: Kluwer Academic Publishers.
Gasset, Jose Ortega Y (1941). "Man, the Technician," Chapter 3. *Toward a Philosophy of History*. New York: W. W. Norton.
Gewirth, Alan. (1985). "Rights and Virtues." Review of Metaphysics, 38: 739–62.
Gewirth, Alan (1998). *Self-fulfillment*. Princeton: Princeton University Press.
Gewirth, Alan. (1996). *The Community of Rights*. Chicago: University of Chicago Press.
Gewirth, Alan. (1978). *Reason and Morality*. Chicago: University of Chicago Press.
Giubilini, Alberto and Savulscu, Julian. (2018). "The Artificial Moral Advisor - The Ideal Observer Meets Artificial Intelligence." *Philosophy and Technology* 31: 169–188.
Goldberg, Elkhonon. (2005). *The Wisdom Paradox: How Your Mind Can Grow Stronger as Your Brain Grows Older*. UK: The Free Press.

Good, Irving John. (1965). "Speculations Concerning the First Ultraintelligent Machine." *Advances in Computers,* 6: 31–83.

Grau, Christopher and Smuts, Aaron. (2017). *Oxford Handbook of the Philosophy of Love.* Oxford:Oxford University Press.

Hadot, Pierre. (2001). The Inner Citadel: The Meditations of Marcus Aurelius. Cambridge MA: Harvard University Press.

Hall S. Stephen (2010). *Wisdom: From Philosophy to Neuroscience*: Brisbane: University of Queensland Press.

Hadot, Pierre (2006). *The Veil of Isis.* Translated by Michael Chase. Cambridge Massachusetts: The Belknap Press of Harvard University Press.

Hadot, Pierre (1995). *Philosophy as a Way of Life.* Oxford: Blackwell Publishing.

Harris, Robert. (2012). *The Fear Index.* London: Vintage.

Harris, Robert. (2019). *The Second Sleep.* London: Hutchinson.

Heersmink, Richard. (2017). "Distributed Cognition and Distributed Morality: Morality, Agency and Systems." *Science Engineering Ethics,* 22(2): 431–448.

Hui, J. and Spence, Edward H. (2016). Internet Addiction and Well-Being: Daoist and Stoic Reflections. *Dao: A Journal of Contemporary Philosophy.* Springer.

Hunter, Fergus. (2019). "Consumer Watchdog Takes Legal Action Against Google for Misleading Conduct." *The Sydney Morning Herald,* October 29.

International Association for Safe and Ethical AI (IASEAI). (2025). Inaugural Conference Feb 6–7, Paris, France. https://www.iaseai.org/conference.

Jeste, Dilip V. and Lee, Ellen E. (2019). "The Emerging Empirical Science of Wisdom: Definition, Measurement, Neurobiology, Longivity and Interventions." *Harvard Review Psychiatry,* 27(3): 127–140.

Johnson, Jeffery. (2014). "The Ethics of Big Data in Higher Education." *International Review of Information Ethics,* 21(7): 3–10.

Kant Immanuel. (1981). *Grounding for the Metaphysics of Morals.* Indianapolis: Hackett Publishing Company.

Kekes, John. (1995). *Moral Wisdom and Good Lives.* Ithaca: Cornell University Press.

Kekes, John. (1983). "Wisdom," *American Philosophical Quarterly,* 20(3): 277–286.

Kim, Tae Wan, and Mejia, Santiago. (2019). "From Artificial Wisdom: What Socrates Teaches Us". *Computer,* 52(10): 70–74.

Kitchin, Robert. (2016). "Thinking Critically About Researching Algorithms." *Information, Communication and Society,* 20(1): 1–16.

Kunzmann, Ute and Baltes, Paul. (2005). "The Psychology of Wisdom: Theoretical and Empirical Challenges", in Robert J Stenberg and Jennifer Jordan (eds), *A Handbook of Wisdom: Psychological Perspectives.* New York: Cambridge University Press.

Laertius, Diogenes. (1925). *Lives of Eminent Philosophers, Book 6–10.* CM: Harvard University Press.

Lehrer, Tom and Smith, Chris. (1996). *Knowledge, Teaching and Wisdom.* Dordrecht: Kluwer Academic Publishers.

Lessig, Lawrence. (2013). "Institutional Corruption - Defined." *Social Sciences*, 41(3): 553.

Levy, Steven. (2003). "Smarter AI Assistants Could Make it Harder to Stay Human". *Wired*, Sept 22, 2023.

Long, A.A. (1986). *Hellenistic Philosophy – Stoics, Epicureans, Sceptics*. 2nd Edition. Berkeley: University of California Press.

Lovelock, James. (2020). *Novacene*. Penguin, Random House UK.

Lynch, Michael P. (2016). *The Internet of Us: Knowing and Understanding Less in the Age of Big Data*. London: Penguin, Random House.

Lyons, Jonathan. (2010). *The House of Wisdom: How the Arabs Transformed Western Civilisation*. UK: Bloomsbury.

Manders-Huit, Naomi and Van den Hoven, Jeroen. (2008). "Moral Identification in Identity Management Systems." *International Federation for Information Processing Digital Library*, https://doi.org/10.1007/978-0-387-79026-8_6.

Mascaro, Juan. (1962). *Bhagavad Gita*. Translated from the Sanskrit by Juan Mascaro. Harmondsworth: Penguin Classics.

Matthias, Andreas. (2004). "The Responsibility Gap: Ascribing Responsibility for the Actions of Learning Automate." *Ethics Information Technology*, 6: 175–183.

Maxwell, Nicholas. (1984). "From Knowledge to Wisdom: Guiding Choices in Scientific Research." *Bulletin of Science, Technology and Society*, 4(4): 316–334.

Maxwell, N. (2007). *From Knowledge to Wisdom* (2nd Ed). London:Pentire Press.

McEwan, Ian. (2019). *Machines Like Me and People Like You*. London: Jonathan Cape.

Milkman, Oliver (2019), "Defiant Mark Zuckerberg Defends Facebook Policy to Allow False Ads." *The Guardian*, December 2, 2019.

Miller, Seamus. (2010). *The Moral Foundations of Social Institutions*. Cambridge University Press.

Miller, Seamus. (2018). "Corruption". *Stanford Encyclopedia of Philosophy* (Winter 2018 Edition), Edward N Zalla (ed).

Miller, Seumas and Bossomaier, Terry. (2024). *Cybersecurity Ethics, and Collective Responsibility*. Oxford: Oxford University Pres.

Miller, Seamus., Roberts, Peter and Spence Edward H. (2005). *Corruption and Anti-corruption: An Applied Philosophical Approach*. Upper Saddle River NJ: Pearson/Prentice hall.

Mims, Christopher. (2018). "Who Has More of Your Personal Date Than Facebook?" *The Australian and Wall Street Journal*, April 23.

Mittelstadt, Brent et al. (2016). "The Ethics of Algorithms: Mapping the Debate." *Big Data and Society*, 3(2): 68

Mittelstadt, Brent., Allo, Patrick and Floridi, Luciano. (2016). "The ethics of algorithms: Mapping the debate." *Bid Data and Society*. July–December: 1–21.

Mittelstadt, Brent. D., Allo Patrick., Taddeo, Mariarosaria., Wachter, Sandra., and Floridi Luciano. (2016). "The Ethics of Algorithms: Mapping the Debate." *Big Data and Society*. July–December: 1–21.

Nozick, R. (1989). "What is Wisdom and Why Do Philosophers Love It?" *The Examined Life*. New York: Touchstone Press.
Nussbaum, Martha. (1994). *The Therapy of Desire: Theory and Practice in Hellenistic Ethics*. Princeton, New Jersey: Princeton University Press.
Nussbaum, Martha. (1997). *Cultivating Humanity: A Classical Defence of Reform in Liberal Education* Cambridge, Massachusetts, and London: Harvard University Press.
Nussbaum, Martha. (2000)."Duties of Justice, Duties of Material Aid: Cicero's Problematic Legacy." In *Symposium on Cosmopolitanism, Journal of Political Philosophy*, 8(2): 191
O'Neil, Cathy. (2016). *Weapons of Math Destruction: How Big Data Increases Inequality and Threatens Democracy*. New York: Penguin Books.
O'Neill, Onora. (2013). "What We Don't Understand About Trust." *TED Talk, Houses of Parliament England*. https://www.ted.com/talks/onora o neill.
Ord, Toby. (2020). *The Precipice: Existential Risk and the Future of Humanity*. London and Dublin: Bloomsbury.
Orlowski, Jeff. (2019). The Social Dilemma. Netflix Documentary
Pasquale, Frank. (2015). *The Black Box Society: The Secret Algorithms That Control Money and Information*. Cambridge, Massachusetts: Harvard University Press
Plato. (1952). *The Apology of Socrates*. 2nd Edition. Penguin Classics.
Plato. (1978). "The Apology" in The Collected Works of Plato. Edith Hamilton and Huntington Cairns (eds) Princeton: Princeton University Press.
Plato. (1997). *Plato: Symposium and Death of Socrates*. Translated by Tom Griffith. Hertfordshire: Wordsworth Classics.
Plato. (2007). *The Republic*. Penguin Classics.
Robson, David. (2020). *The Intelligence Trap: Revolutionise Your Thinking and Make Wiser Decisions*. UK: Hodder & Stoughton.
Rughinis, Cosmina, Rughinis, Razvan, and Mateni, Stefina. (2015). "A Touching Voice App Thinking About Ethics of Persuasive Technology Through an Analysis of Smoking Cessation App." *Ethics and Information Technology*, 17(4): 1–15.
Russel, Stuart. (2019). *Human Compatible*. UK: Penguin Books.
Ryan, Sharon. (1996). "The Epistemic Virtues of Consistency." *Synthese*, 109(2): 121–141.
Ryan, Sharon. (2014) "Wisdom." *Stanford Encyclopedia of Philosophy*. Accessed May 30, 2025.
Ryan, Sharon. (1999). "What is Wisdom?" *Philosophical Studies*, 93: 119–139.
Seligman, Martin. (1991). *Learned Optimism: How to Change your Mind and Your Life*. New York: Vintage Books.
Sevilla, David. (2013). "The Quest for Artificial Wisdom." *AI and Society*, 28(2): 199–207.

Schneider, Susan. (2019). *Artificial You: AI And the Future of Your Mind.* Princeton and Oxford: Princeton University Press.

Sherman, Nancy. (2021). *Stoic Wisdom: Ancient Lessons for Modern Resilience.* Oxford University Press.

Sinha, A and Lakhanpal, P. (2024). "Can AI Systems Become Wise? A Note on Artificial Wisdom." *AI & Society*, 39: 2153–2154.

Smith, Justine E.H. (2022). *The Internet Is Not What You Think IT IS: A History, A Philosophy, A warning.* Princeton and Oxford: Princeton University Press.

Spence, Edward H. (1999). "A Commentary on Deryck Beyleveld's - Gewirth and Kant on Justifying the Supreme Principle of Morality". In Michael Boylan (ed) *Gewirth, Critical Essays on Action, Rationality, and Community.* Maryland: Rowman and Littlefield Publishers.

Spence, Edward H. (2004). "Philosophy Plays: A neo-Stoic Method for Teaching Ethics". *Teaching Ethics*, 5(1): 41–57.

Spence, Edward H. (2006). *Ethics Within Reason: A Neo-Gewirthian Approach.* Lanham: Lexington Book.

Spence, Edward H. (2006). "Corruption Doomed to Fail in Long Term." *Financial Review*, January 14.

Spence, Edward H. (2007). "Positive Rights and the Cosmopolitan Community: Right-Centered Foundations for Global Ethics." Journal of Global Ethics, 3(2):July/August.

Spence, Edward H. (2008). "Corruption in the Media." *International Journal of Applied Philosophy*, 22(2): 231–241.

Spence, Edward H. (2009). "A Universal Model for the Normative Evaluation of Internet Information." *Ethics and Information Technology*, 11(4): 243–253.

Spence, Edward, H. (2011a). "The Theatre of Philosophy: A Neo–Socratic Method for Performing Public Philosophy', in Bull, Gordon (Ed.), *Creativity: Brain – Mind – Body: A view into the future of Australian Art and Design Schools, the 2011 Australian Council of University Art and Design Schools (ACUADS) Conference*, the Australian National University, the University of Canberra, and the Canberra Institute of Technology, 21 – 23 September 2011.

Spence, Edward H. (2011b). "Is Technology Good for Us? A Eudaimonic Meta-Model for Evaluating the Contributive Competence of Technologies for a Good Life." *Nanoethics*, 5: 335–343.

Spence, Edward H. (2011c). "Information, Knowledge, and Wisdom: Groundwork for the Normative Evaluation of Digital Information and Its Relation to the Good Life", *Ethics and Information Technology*, 13(3): 261–275. https://doi.org/10.1007/s10676-011-9265-7

Spence, Edward H. (2013). "Wisdom and Well-being in a Technological Age," *Synesis: A Journal of Science, Technology, Ethics, and Policy* 2013

Spence, Edward H. (2014). "The Advertising of Happiness and the Branding of Values." In Michael Boylan (ed). *Business Ethics* (2nd Ed), Upper saddle River NJ:Pearson/Prentice Hall.

Spence, Edward H. (2016). "Is Alan Gewirth's Moral Philosophy Neo–Stoic?" In Per Bauhn (Ed.) *Gewirthian Perspectives on Human Rights.* New York: Routledge.

Spence, Edward H. (2017). "Corruption in the Media" in Michael Ablander and Sarah Hudson (eds) *The Handbook of Business and Corruption: Cross Sectional Experiences.* Emerald Group Publishing.

Spence, Edward H. (2018) "Harm in Media Marketing: The Branding of Values", *Communication and Media Ethics,* Patrick Plaisance (ed), Berlin, Boston:de Gruyter Mouton.

Spence, Edward H. (2020). "The Sixth Estate: Tech Media Corruption in the Age of Information." *Journal of Information, Communication and Ethics in Society,* 18(4): 553–573.

Spence, H. Edward. (2021a). *Stoic Philosophy and the Control Problem of AI Technology: Caught in the Web.* Rowman and Littlefield.

Spence, H. Edward. (2021b). *Media Corruption in the Age of Information.* Switzerland: Springer Nature.

Spence, Edward H. and Van Heekeren, Brett. (2005). *Advertising Ethics,* New Jersey:Pearson Prentice Hall.

Spence, Edward H., Alexandra Andrew., Quinn, Aaron. and Dunn, Anne. (2011). *Media, Markets and Morals.* Oxford: Wiley-Blackwell.

Stohr, Karen. (2022). *Choosing Freedom: A Kantian Guide to Life.* UK: Oxford University Press.

Striker, Gisela. (1992). "Stoicism" in Lawrence C. Becker and Charlotte B. Becker, Editors, *Encyclopedia of Ethics.* New York: Garland Publishing.

Striker, Gisela. (1996). *Essays On Hellenistic Epistemology and Ethics.* Cambridge: Cambridge University Press.

Suleyman, Mustafa and Bhaskar, Michael. (2023). *The Coming Wave: Technology, Power and the Twenty-First Century's Greatest Dilemma.* New York: Crown Publishing Group.

Swartwood, Jason and Tiberius, Valerie. (2019). "Philosophical Foundations of Wisdom" In *The Cambridge Handbook of Wisdom,* (Eds Robert Sternberg and Judith Glueck) Cambridge University Press.

Tegmark, Max. (2018). *Life 3.0: Bring Human in the age of Artificial Intelligence.* UK: Penguin Books.

Tiberius, V. (2008). *The Reflective Life: Living Wisely Within Our Limits.* Oxford:Oxford University Press.

Tsai, Cheng-Hung. (2020). "Artificial Wisdom: A Philosophical Framework." AI & Society. https://doi.org/10.1007/s00146-020-00949-5

Tufekci, Zeynep. (2018). Facebook's Surveillance Machine, *The New York Times,* March 19, 2018.

Tufekci, Zeynep. (2018). "You Tube, The Great Radicalizer." *The New York Times,* March 10.

Turing, Alan. (1951). "Can Digital Computers Think?" *BBC Broadcast*, 15th May 1951.

Umbrello, Steven and Van de Poel, Ibo. (2020). "Mapping Value Sensitive Design Onto AI for Social Good Principles." *AI and Ethics*. https://doi.org/10.1007/s43681-021-00038-3.

Vallor, S. (2016). *Technology and the Virtues: A Philosophical Guide to the Future Worth Wanting*. Oxford: Oxford University Press.

Van Dijck, Jose, Poell Thomas, and De Waal, Matijn. (2018). *The Platform Society: Public Values in a Connective World*. Oxford: Oxford University Press.

Varelius, J. (2004). "Objective Explanations of Individual Well-Being." *Journal of Happiness Studies*, 5: 73–91.

Véliz, Carissa. (2020). *Privacy is Power: Why and How You Should Take Back Control of Your Data*. London: Bantam Press.

Verbeek, Peter-Paul. (2011). *Moralizing Technology: Understanding and Designing Moral Things*. Chicago: Chicago University Press.

Verbeek, Peter-Paul. (2014). *The Moral Status of Technical Artifacts*. Dordrecht: Springer.

Vitec, B and Jackson, W. (2008). *The Virtues of Ignorance: Complexity and Sustainability, and the Limits of Knowledge*. Lexington, Kentucky: The University Press of Kentucky.

Wachter, Sandra and Mittlestast, Brent. (2019). "A Right to Reasonable Inferences: Rethinking Data Protection Law in the Aid of Big Data and AI." *Columbia Business Law Review*, 1: 1–130.

Walsh, Toby. (2022). *Machines Behaving Badly: The Morality of AI*. Collingwood Australia: La Trobe University Press.

Webb, Amy. (2019). *The Big Nine: How the Tech Titans and Their Thinking Machines Could Warp Humanity*. New York: Public Affairs, Hatchett Book Group.

Wierner, Norbert. (1960). "Some Moral and Technical Consequences of Automation." *Science*, 131(3410): 1335–1358.

Zarsky, Tal. (2016). "The Trouble with Algorithmic Decisions: An Analytical Road Map to Examine Efficiency and Fairness in Automated and Opaque Decision Making." *Science, Technology and Human Values*, 41(1): 118–132.

Zizek, Slavoj. (2018). *Like A Thief In Broad Daylight: Power in the Era of Post-Humanity*. UK: Penguin Books.

Zuboff, Shoshana. (2019). *The Age of Surveillance Capitalism: The Fight for a Human Future at the New Frontier of Power*. London: Profile Books Ltd.

Zuckerberg, Mark. (2019). "Four Ideas to Regulate the Internet." *Facebook*, March 30, 2019.

Zuckerberg, Mark. (2020). "Big Tech Needs More Regulation." *The Financial Times*. February 16, 2020.

Index

E. H. Spence, *Wisdom in the Age of Intelligent Machines*,
https://doi.org/10.1007/978-3-031-93883-2

The manufacturer's authorised representative in the EU is Springer Nature Customer Service Centre GmbH, Europaplatz 3, 69115 Heidelberg, Germany. If you have any concerns regarding our products, please contact ProductSafety@springernature.com

Printed and bound by CPI Group (UK) Ltd, Croydon, CR0 4YY
01/10/2026
02225967-0009